# SNAKES OF
# AUSTRALIA

## DANGEROUS & HARMLESS

**Peter Mirtschin & Richard Davis**

HILL OF CONTENT
Melbourne

First published in Australia 1992
by Hill of Content Publishing
86 Bourke Street, Melbourne 3000
Reprinted 1995, 1998, 2000

Cover design: Julian Jones
Cover photograph: Common tiger snake *Notechis scutatus*
by Peter Mirtschin
Typeset in Australia by Abb-type, Collingwood Victoria
Printed and bound in Singapore
by Kyodo Printing Co (Singapore) Pte Ltd

National Library of Australia
Cataloguing-in-Publication data

Mirtschin, Peter, 1947–
  Snakes of Australia: dangerous and harmless.

  Includes index.
  ISBN 0 85572 209 6.

  1. Snakes — Australia — Identification. I. Davis, Richard,
  1944–  . II. Title.

597.960994

# Contents

# Acknowledgments

It would be impossible to list all the people who have helped to compile this field guide of Australian snakes. For those who have contributed, we are most grateful. Special thanks to Steve Wilson, Greg Harold and Hal Cogger, without whose photography this book would not have been possible. Other photographers who helped to fill gaps where needed were Paul Horner, Lyal Naylor, John Weigel, David Robinson and Chris Banks. We are indebted to the many people who supplied advice and their papers on various aspects of snakes — Rick Shine, Jeanette Covacevich, Terry Schwaner and Hal Heatwole. We are also grateful to the Australian National Parks and Wildlife Service for information on endangered species. Our deepest thanks to Phil Parks and Wayne Court for their assistance while the manuscript was being prepared, and to Struan Sutherland, who has always been willing to freely supply information on venoms and toxicity of the various venomous snakes.

Thanks to M. Schubert for scalation drawings.

# Introduction

Snakes are a fascinating part of Australia's fauna. Even people who don't like snakes often are intrigued with this unique part of Australian wildlife that has some of the largest pythons and deadliest snakes in the world. The types included in our fauna are pythons, file snakes, blind snakes, rear-fanged or solid-toothed harmless snakes, sea snakes and the elapids or cobra-like snakes.

This field guide is designed to assist people in identifying snakes and also to provide a summary of all the available information about these species. Hopefully, it will stimulate an interest in a section of our fauna that is often overlooked when considering conservation measures and will encourage an attitude of coexistence with some of our 'deadly species' that do not present anywhere near the risk often attributed to them.

In reading through this book, you will soon find that there are many gaps in our knowledge. You can help fill in those gaps by noting your observations and passing them on to your nearest state museum. Museums are places where much backroom research is carried out, and those researchers are always interested in any observation you might like to offer.

There are seven terrestrial and five marine snake types that we consider to be threatened. Four terrestrial and all of the marine species are considered threatened only because a few specimens have ever been collected. However, if some of the environmental pressures continue, others will follow. Taking care of the environment and helping to rehabilitate habitats as much as possible all helps to reverse the trend that has crippled some environments in Australia.

We hope you find this book useful and that it encourages you to care for our ophidian fauna, just as we do for our koalas and kangaroos. Take care in the bush and enjoy, at a distance, the wonderful world of Australian snakes.

# Snakes and people in Australia

Australia is rich and diverse in its snake fauna. There are from 175 to 190 recognised species and subspecies of terrestrial and sea snakes on the mainland, its islands and surrounding waters. The deadliest snakes in the world inhabit Australia. Around 22 Australian snake types are either as deadly as or more deadly than the Asian cobra *Naja naja*, and 16 of these are much more deadly. Probably the deadliest snake in world, the inland taipan *Oxyuranus microlepidotus*, which in relative terms has only recently been rediscovered and found to be 50 times more deadly than the cobra, inhabits one of the harshest environments for reptiles in the world.

Australia has 15 python types,(including the smallest python *Liasis perthensis* and one of the largest pythons *Morelia amethistina*), 28 blind snakes, 32 sea snakes (all of which must be considered dangerous), 2 file snakes and 10 rear-fanged or harmless snakes. Its largest ophidian component is represented by the elapids or front-fanged, cobra-like snakes, of which there are about 90 different types. It is this group of snakes by which Australia is best known and of which 30 types are dangerous to humans.

Since early settlement of Australia, snakes have been used for a wide variety of reasons. The snakemen of the side shows were always popular and even today the modern day snakemen are similar in many ways. They are special people, who are considered by some, to have a weird twist in their outlook, in that that they seek pleasure in displaying animals that are frightening and very deadly. Nevertheless, people flock to see these snakemen, perhaps hoping to witness a mishap when the unfortunate snake handler gets bitten. Other more refined snake presentations are to be found at zoos, reptile parks and nature displays, where the educational intention is far more evident.

Snakes have been used for skins, and sea snakes are still being harvested for this reason. The garments such as belts, wallets and shoes that have been made from snake skins could be described as attractive. The practice of harvesting snakes for this reason would be questioned by some; however, if there is no threat to their populations, then consideration should be given to an industry of this type under a satisfactorily controlled plan. Snakes have been used for food and some species are still used to a limited extent by Aborigines as food.

Snakes are valuable components of our fauna in the tourism industry. The attractions that draw tourists are forever changing as they search for new experiences. It is not unforeseeable that in the future expeditions will be offered to some overseas visitors to venture into the outback to see some of the unusual Australian ophidian fauna. For the same reasons that crocodiles have become a huge tourist attraction in the Northern Territory, it is possible that snakes will also offer the same opportunities to some entrepreneurial tourist operators.

Snake venoms have been used for the manufacture of antivenom for the treatment of snake bite since 1929 in Australia. Since then considerable research has been carried out with Australian snake venoms both in Australia and overseas. The understanding of the human (and other animal) neuromuscular systems has been studied using neurotoxins from a number of Australian snake venoms, and along with the Asian many-banded krait *Bungarus multicinctus*, the medical profession owes much of that knowledge to the Australian tiger snake *Notechis scutatus*, the taipan *Oxyuranus scutellatus* and more recently the brown snake *Pseudonaja textilis*. While the disease called myasthenia gravis can be diagnosed using the many-banded krait venom *B. multicinctus*, the Western Australian dugite *Pseudonaja affinis* venom can also be used for that purpose.

Australian snake venoms are a rich source of blood-clotting components. Although often causing the same result, they all have their own peculiarities and each one of them potentially has a use in medicine both as blood-clotting assays and in the therapeutic area.

As well as blood clotters, anticoagulants are present in a number of venoms and again these are proving to be exciting in medical research. It is feasible that some anticoagulants from the king brown snake *Pseudechis australis* could hold the answer to many problems associated with the current anticoagulants used in medicine. Some of the anticoagulants (there could be 10) in the king brown snake venom are very close to some of the naturally occurring human enzymes.

The other main toxin in a number of venoms are the mytoxins or muscle-destroying components. These components could be used in studying diseases such as muscular distrophy and other neuromuscular diseases. Not only do they produce symptoms in animals that sometimes resemble muscle diseases, but it is possible that the specific antibodies against these myotoxins could be useful in treating certain types of diseases.

Snakes as pets are becoming more popular. With the enormous numbers of dogs and cats and other exotic animals that are kept by Australian people, if some could alternatively keep native animals such as snakes, our environment would be better protected against the influx of unwanted exotica through releases or escapes etc. In some cases, further contributions will add to our existing knowledge concerning those species and there would be a spin-off advantage in winning other people over to take an interest in wildlife.

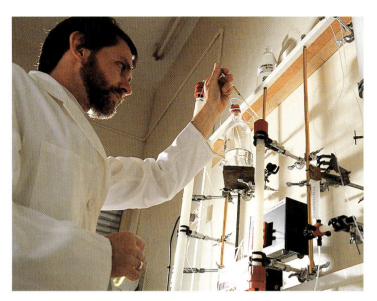

*Scientist Phil Mugg separating neurotoxins from snake venom at the Institute of Medical and Veterinary Science Adelaide. (photo by M. Fitzgerald).*

*Extracting venom at Venom Supplies Laboratory, Tanunda, South Australia. (photo by C.D. Van Meurs).*

# Identification

In using the book, we have excluded complicated keys and ask readers to use the photographs as a start in identification, followed by the descriptions of the snakes. Our system is not foolproof and, to be done correctly, many of the colour variations would need to be included. Space does not permit us to include the many colour variations of some of the species covered in the book.

It goes without saying that, when identifying snakes, all care should be taken when the deadly species are involved. These species are better left alone unless they are dead, and we don't advocate killing them just for identification. However, if the snake is dead, such as a roadkill, it can be useful to try to identify it because it is always possible new species or further range extensions will be found.

The museums in the respective states are always grateful for roadkill specimens, however permits from the National Parks and Wildlife Services are required in most cases for specimen collection. Specimens can be preserved in formalin (10:1 water/formalin) or in spirit. Freezing specimens is also a good way of temporarily keeping specimens until they can be correctly preserved.

The scale counting technique is shown in the following diagrams and the names of the various scales are labelled.

## HEAD SHIELDS

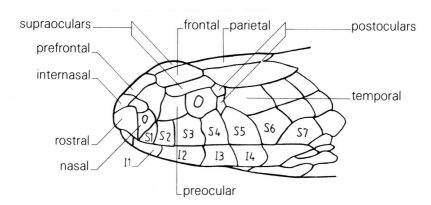

Illus. by M. Schubert

I1–I4 = infralabials
S1–S7 = supralabials

12

# BELLY SCALES

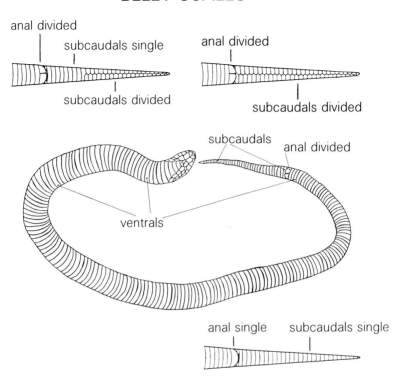

anal divided

subcaudals single

subcaudals divided

anal divided

subcaudals divided

subcaudals

anal divided

ventrals

anal single    subcaudals single

# MIDBODY SCALES

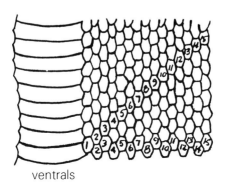

ventrals

# Pythons
## *Family BOIDAE*

Pythons are the largest of the Australian snakes. The family also includes possibly the world's smallest python, which occurs in Western Australia. They are usually bulky and slow moving, and feed mainly on birds and mammals. However, the *Aspidites* genus also feed on other reptiles. They are all egg layers and coil around the eggs to 'incubate' them. By continually moving in what appears to be a shivering action, while coiled around the eggs, pythons are able to slightly elevate the temperature of the egg clutch by friction.

Most pythons are nocturnal, but can be found basking during the day. They are all non-venomous, possessing only sharp recurved (backwards-curving) teeth, with no fangs or venom glands. They kill their prey mostly by constriction — coiling around the prey and tightening their coils as the animal exhales, effectively killing it by asphyxiation. *Aspidites ramsayi* has been observed using an adaptation of this method by pressing its prey against its enclosure walls with its body.

All pythons have cloacal spurs (vestigial limbs) on either side of the anal scale, and they are the only snakes to have a pelvis.

With the exception of the *Aspidites* genus, they all have a number of heat-sensing pits in some of their lower labial scales, and sometimes in some of the upper labial scales and rostral scales. These are used to seek out warm-blooded prey.

Pythons usually have prehensile tails.

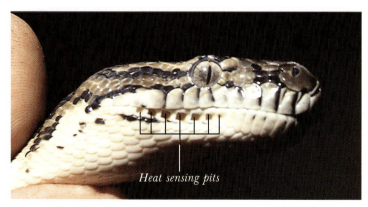

*Heat sensing pits*

*Carpet python* Morelia spilotes variegata *showing pitted labial scales. (photo by P. Mirtschin).*

*Heat sensing pits*

*Carpet python* Morelia spilotes variegata *showing pitted rostral scales. (photo by P. Mirtschin).*

*A woma python* Aspidites ramsayi *showing the lack of pitted labial scales (photo by P. Mirtschin).*

*Cloacal spurs*

*The underside of a carpet python* Morelia spilotes variegata *showing the cloacal spurs.*

# Black-headed python

*Aspidites melanocephalus* (Krefft, 1864)
*(Non-venomous)*

**Description** Head slightly pointed and only slightly distinct from a robust body. Grows to 3.0 m in length. Dorsally, light to dark straw-yellow with darker brown colours, forming in most cases a series of irregular-shaped bands. The head, nape and throat are a glossy jet black. The ventral surface is creamish with darker blotches, representing a continuation of the dorsal bands. Juveniles are more vividly marked. Has a cloacal spur on either side of the vent. Midbody scales are in 50–65 rows, ventrals number 315–359, subcaudals 60–75 (mostly single) and the anal scale is single. Lacks heat-sensing pits on the infralabial scales.

**Habits and habitat** Occurs in subhumid to humid northern Australia (not in extremely arid parts), where it seeks refuge in crevices, caves, hollow logs and abandoned burrows. Mainly nocturnal and sometimes diurnal on cool days, particularly after rain. Like all pythons, it is oviparous and produces 8–11 eggs. Feeds on mammals, birds and reptiles, including venomous snakes. Has a threatening disposition when disturbed, displayed by elevating its head and hissing. One suggestion for the black head is that it can allow heat absorption whilst exposing only a small portion of its body.

# Woma or Ramsay's python

*Aspidites ramsayi* (Macleay, 1882)
*(Non-venomous)*

**Description** Slightly pointed head is indistinct from a robust body. Grows to 3.0 m in length. Dorsally, dull straw-yellow with either dull olive-brown bands or distinct orange-brown bands, which often diffuse in the centre to form a darker area along the vertebral line. The head is either a bright orange or yellow or the same as the ground colour of the body. The ventral surface is cream to yellow. Has a cloacal spur on either side of the vent. Midbody scales are in 50–65 rows, ventrals number 280–315, subcaudals 45–55 (mostly single) and the anal scale is divided. Lacks heat-sensing pits on the infralabials.

**Habits and habitat** Nocturnal but sometimes diurnal. Found over a wide range of habitat types, including arid deserts, rocky ranges, subhumid environments, flood plains, sand dunes supporting hummock grasses, woodlands, shrublands and heathlands. Seeks refuge mainly in abandoned animal burrows. It is oviparous and lays up to 28 eggs. Feeds on mammals, reptiles (including venomous snakes, especially *Pseudonaja nuchalis*) and birds.

In captivity, woma pythons have been observed pressing their prey against the sides of their hide boxes. In fact, some specimens have refused to feed until they were given food in confined spaces, enabling them to carry out this unusual prey catching and killing procedure. It is assumed that this technique is used by the species when pursuing prey in burrows. This proposal is supported by the large number of individuals with scars and defects along their bodies and tails. One expert has also suggested that woma pythons use their tails as a lure. One of the authors (Peter Mirtschin) observed a woma python stalking dingo pups (*Canis familiaris dingo*) on Goyder's Lagoon in northern South Australia. There is also an old record of a woma python devouring a small wallaby.

# Green python

*Chondropython viridis* (Schlegel, 1872)
*(Non-venomous)*

**Description** Head large and distinct from a robust body. The body is roughly triangular in cross-section. It has a prehensile tail. Grows to 2.0 m in length. Dorsally, any shade of emerald green, with a partially broken vertebral row of white scales running the full length of the vertebral line (sometimes discontinuous), with regular blue or yellow scales abutting on either side of the vertebral line. Narrow transverse blue lines are occasionally present on the sides. The ventral surface is creamish to yellowish. Juveniles can be either golden-yellow, orange, brown to red, with white markings on the sides and vertebral line, and a purplish streak through the eyes. The different colours in juveniles can last up to three years and the transition to adult colours takes about three weeks. It has some pitted infralabial scales and a cloacal spur on either side of the vent. Midbody scales are in 50–75 rows, ventrals number 225–260, sub-caudals 90–110 (single) and the anal scale is divided.

**Habits and habitat** An arboreal species, found in trees, shrubs and bamboo thickets, its colour providing an effective camouflage against predators and prey. Nocturnal and restricted to rainforests and bamboo thickets of northern Queensland. Its characteristic posture is one of symmetric coils over a branch, with its head positioned in the middle. It is oviparous and produces 11–25 eggs. Feeds on birds and mammals. Has been observed vibrating its thin bluish tail to attract prey. Drinks water collected in its body coils. Superficially, very similar to another species from South America, the emerald tree boa *Corallus caninus*.

# D'albertis's python

*Liasis (Bothrochilus) albertisii* (Peters & Doria, 1878)
*(Non-venomous)*

**Description**  Head distinct from a robust body. Grows to 3.0 m in length. Dorsally, uniform grey-brown, the head being slightly darker. The labial scales are creamish, edged with black. The ventral surface is creamy white. Some of the infralabials are pitted. Has a cloacal spur on either side of the vent. Midbody scales are in 45–55 rows, ventrals number 260–290, subcaudals 60–80 (divided) and the anal scale is single.

**Habits and habitat**  A nocturnal forest dweller. Occurs on some of the Torres Strait islands near New Guinea, where it is found in monsoon forests, regrowth areas and coastal palm swamps. It is an egg layer and has been known to produce 30 eggs. Feeds on mammals and birds.

# Children's python

*Liasis (Bothrochilus) childreni* (Gray, 1842)
*(Non-venomous)*

**Description** A number of different populations of this species are thought by some taxonomists to be distinct species and, as such, are separated into *Bothrochilus childreni*, *Bothrochilus maculosus* and *Bothrochilus stimsoni* (two subspecies). In this treatment they are all included in *Liasis (Bothrochilus) childreni*.

Head distinct from a robust body. Grows to 1.5 m in length. Anterior labials not deeply pitted. Possesses posterior pitted infra-labial scales. Has a cloacal spur on either side of the vent. Dorsally, whitish, creamish or olive, with numerous darker spots (sometimes arranged in groups), which appear like bands over the entire length of the body. Some specimens are a uniform colour without spots. The ventral surface is white to cream. Midbody scales are in 31–49 rows, ventrals number 243–302, subcaudals 30–45 (mostly divided) and the anal scale is single.

**Habits and habitat** Nocturnal. Occurs over a wide range of habitats, from inland deserts, flood plains, rainforests and semi-arid woodlands, where it seeks refuge in rocky outcrop crevices, hollow limbs, abandoned burrows, termite mounds, surface debris, flood plain cracks and caves. It is an egg layer and produces 4–20 eggs. Feeds on lizards, mammals and birds.

# Water python

*Liasis (Bothrochilus) fuscus* (Peters, 1873)
*(Non-venomous)*

**Description** Head distinct from a robust body. Grows to 3.0 m in length. Dorsally, light to dark iridescent olive-green, becoming lighter in colour on the flanks, where it diffuses into a bright or dull yellow on the ventral surface. The throat is creamy white. The anterior supralabials are pitted, as are some of the posterior infralabials. The underside of the tail is a dark olive to black. Has a cloacal spur on either side of the vent. Midbody scales are in 40–55 rows, ventrals number 270–300, subcaudals 60–90 (divided) and the anal scale is single.

**Habits and habitat** Nocturnal. Usually found in or close to water — from freshwater swamps, lagoons, river margins, creeks, dams and reservoirs. Feeds mainly on water birds and sometimes their eggs, but has been recorded eating mammals and small crocodiles. During the day, it shelters beneath rocks, in logs, stumps and mats of dead vegetation. It is an egg layer and produces clutches of 9–22 eggs.

# Olive python

*Liasis (Bothrochilus) olivaceous olivaceous* (Gray, 1842)
*(Non-venomous)*

**Description** Head distinct from a robust body. Grows to 3.7 m in length. Dorsally, light to dark olive-brown, fading on the flanks to colour of the ventrals. The lips, side of throat, neck and ventral surface are white or cream. The lips may be finely dotted with darker pigment. The anterior labials and infralabials are pitted. Has a cloacal spur on either side of the vent. Midbody scales are in 58–72 rows, ventrals number 321–411, subcaudals 90–119 (divided) and the anal scale is single.

**Habits and habitat** Nocturnal and sometimes diurnal. Found in mountain ranges, savannah woodlands, along watercourses and gorges in the arid to humid areas of its range. It is an egg layer and 11 eggs have been recorded. Feeds on mammals (often as large as wallabies), birds and reptiles. Has a habit of waiting in ambush on mammal pads.

**Subspecies**

### *Liasis (Bothrochilus) olivaceous barroni* (Smith, 1981)
Differs from *Liasis olivaceous olivaceous* in having less midbody scale rows (58–63 vs 61–72) and more ventral scales (374–411 vs 355–377). Found only in the Pilbara region of Western Australia. Grows to a greater length than *Liasis olivaceous olivaceous*, where specimens of 5.5 m and 6.5 m have been reported.

# Pygmy or western Children's python

*Liasis (Bothrochilus) perthensis* (Stull, 1932)
*(Non-venomous)*

**Description** Australia's, and possibly the world's, smallest python, growing to only 0.6 m in length. Head distinct from a robust body. Dorsally, light brown to reddish brown with numerous darker spots, sometimes arranged to give the appearance of banding. Obscure dark streaks extend from the eye to the angle of the mouth. The ventral surface is creamish or whitish. Midbody scales are in 31–35 rows, ventrals number 205–255, subcaudals 30–45 (mostly divided) and the anal scale is single.

**Habits and habitat** Nocturnal. Found in rocky outcrops, *Acacia*-dominated woodlands and shrublands, and hard stony soils, where it seeks refuge in rock crevices, soil cracks, termite mounds and abandoned burrows. It is an egg layer. Feeds on small lizards, especially geckoes from the *Gehyra* genus.

# Amethystine or scrub python

*Morelia amethistina* (Schneider, 1801)
*(Non-venomous)*

**Description** Elongated head is distinct from a long slender body. It has a prehensile tail. Australia's largest snake, growing to 8.5 m in length, although none have been found this long for many years. A maximum length of between 3.5 m and 7.5 m is probably more normal. Dorsally, light to dark brown, with numerous black spots arranged in a zigzag pattern so that they appear like bands. Specimens from open forests are often lighter in appearance than those from rainforests. When light is reflected from the dorsal skin, there is a brilliant amethyst sheen, giving rise to the species and the common name. The ventral surface is white or cream. Anterior supralabials are deeply pitted. Midbody scales are in 35–50 rows, ventrals number 270–340, subcaudals 80–120 (mostly divided) and the anal scale is single.

**Habits and habitat** Found in open savannah woodlands to rainforests, mangrove swamps, where it shelters in hollow timber, caves, vine thickets and abandoned buildings. Nocturnal, but may also bask during the day in rainforest clearings, especially after a large meal. It is an egg layer and produces 7–20 eggs. Feeds mainly on mammals, even small wallabies. Fruit bats or flying foxes are also taken.

# Centralian carpet or Bredl's python

*Morelia bredli* (Gow, 1981)
*(Non-venomous)*

**Description** Head distinct from a robust body. It has a prehensile tail. Grows to 2.6 m in length. Dorsally, reddish to dark brown, with numerous lighter yellow-tan spots arranged to give the appearance of bands and sometimes oriented in a longitudinal fashion on the flanks. The pattern is more prominent posteriorly. The ventral surface is creamy white, with some posterior ventrals irregularly edged with black. Its anterior supralabials have shallow pits. Midbody scales are in 52–54 rows, ventrals number from 283–295, subcaudals from 82–92 (mostly divided) and the anal scale is single.

**Habits and habitat** An arboreal nocturnal species, found in rocky ranges and outcrops, drainage systems of arid southern NT, where it shelters in hollow trunks and limbs, especially those adjacent to water, rock crevices and caves. It is an egg layer and produces 13–47 eggs. Preys on mammals and birds.

# Rough-scaled python

*Morelia carinata* (Smith, 1981)
*(Non-venomous)*

**Description** Only known from two specimens. Head distinct from a robust body. Grows to 2.0 m in length. Large eyes. Dorsally, pale brown with numerous darker brown blotches, spots and streaks, which are more or less aligned laterally. The ground colour fades and the pattern intensifies posteriorly. The head is more or less without markings, apart from a white diffusion from the eye to behind the head. The lips, chin and ventral surface are cream, with some ventral scales smudged with brown. The dorsal scales are strongly keeled. Midbody scales are in 45 rows, ventrals number 298, subcaudals 83 (mostly divided) and the anal scale is single.

**Habits and habitat** Little is known about this snake, other than that the two specimens are from the Mitchell River Falls on dissected sandstone plateau of Admiralty Gulf, WA.

# Oenpelli rock python

*Morelia oenpelliensis* (Gow, 1977)
*(Non-venomous)*

**Description**  A long head distinct from a long slender body, with a prehensile tail. The largest specimen recorded is 4.57 m; however, there have been suggestions that specimens of 7.0 m may exist. Dorsally, brown to light brown, fading on the flanks, with numerous elongated darker brown blotches aligned longitudinally, being larger dorsally, decreasing in size laterally, fading posteriorly and absent on the tail. Sometimes dorsal scales may be dark edged. A brown streak extends from either the eye or the snout to level with the angle of the jaw. The ventral surface is cream to dull yellow. The anterior supralabials are deeply pitted. Midbody scales are in 70 rows, ventrals number 429–455, subcaudals 155–168 (mostly divided) and the anal scale is single.

**Habits and habitat**  Nocturnal. All specimens have been found near sandstone rock outcrops and gorges of the Arnhem Land Escarpment, where it shelters in deep rock crevices and caves. It is an egg layer and a clutch of 10 eggs has been recorded. The eggs are twice the size of those of *Morelia amethistina*. Feeds on mammals and birds. Some evidence exists that it feeds on Euros *Macropus bernardus*.

# Diamond python and carpet python

*Morelia spilota* (Lacépède, 1804)
*(Non-venomous)*

**Description** Consists of three subspecies, but there is as much variation within the individual subspecies as between them. They all have a large head distinct from a robust body, with a prehensile tail. They grow to 4.0 m in length. The anterior supralabials have shallow pits. Midbody scales are in 40–65 rows, ventrals number 240–310, subcaudals 60–95 (mostly divided) and the anal scale is usually single.

**Habits and habitat** Nocturnally active, but will readily bask during the day. Found over a wide area of Australia in many different habitat types, ranging from rainforests, arid floodplains, along watercourses and channels of the inland drainage systems, rocky mountain ranges, open woodlands, heathlands of the south and offshore islands. Shelter in hollow logs, tree hollows, caves, rock crevices, disused burrows, ceilings of many urban houses (especially in Brisbane). They are egg layers and produce 9–52 eggs in a clutch. Feed on mammals, reptiles and birds. Combat between males has been observed during spring. There is a wide variation in temperament between individuals and, to some extent, populations.

**Subspecies**

***Morelia spilota spilota*** (diamond python)
Dorsally, dark grey to black, with most scales having some cream. Clusters of entirely or almost entirely cream scales forming approximate diamond shapes are located the full length of the body, often in longitudinal rows. The lips are cream and barred with black. The ventral surface is greenish-white, marbled with grey.

29

***Morelia spilota variegata*** (carpet python) (Gray, 1842)
Australia's best-known python. Pale to dark brown, olive-green, grey and reddish, with darker blotches or variegations, sometimes with pale centres. These variegations or blotches sometimes give the appearance of banding or they may be arranged to form longitudinal patterns, especially on the flanks. The ventral surface is greenish-white, marbled with grey.

***Morelia spilota imbricata*** (carpet python) (Smith, 1981)
Has strongly imbricate, lanceolate dorsal scales. Fewer ventrals and fewer subcaudals than *M. s. variegata*. Dorsal colour is brown to blackish-brown, with a variable pattern similar to *M. s. variegata*.

# Front-fanged snakes
## *Family ELAPIDAE*

This group of snakes is well represented in Australia, which has about 44 per cent of the world's total of this genera. They have short, relatively fixed fangs, which are effectively hollow, syringe-like and are formed by the infolding of each side over an anterior groove. The fangs are surrounded by a fleshy sheath called the *vagina dentis* and are connected to a venom duct, which transfers the venom to the fangs from the venom gland.

Most natural habitats have at least one species of elapid. They have two modes of reproduction: ovoviviparous (live bearing) and oviparous (egg laying). The live-bearers are located mainly in the cooler climates.

There is a difference in sizes of the sexes in many species. Where males are larger than females, there has been good correlation to this, indicating that the species exhibits male combat during the breeding season.

About 30 of the elapids are considered dangerous to humans, although some of these would only cause fatalities in extreme conditions. Many Australian elapids are too small to even inflict a bite; however, some are extremely deadly and are by far the most toxic snakes in the world.

# Common death adder

*Acanthophis antarcticus* (Shaw & Nodder, 1802)
*(Highly venomous)*

**Description**  A broad triangular head distinct from the neck, with a short stubby body and a small rat-like tail, terminating in a sharp curved spine. The eyes have elliptical pupils. Specimens of almost 1 m have been recorded, but most average 0.5–0.6 m. Colours are highly variable in any area of its range. Base colour varies from earthy grey to red, with darker transverse crossbands. The tip of the tail is either black, yellowish or white, and is banded in juveniles. The lips have white to cream edges on the dark labial scales, forming bars. Southern death adders usually have black tails, whereas northern specimens are yellow or white. Ventral scales number 110–130, subcaudals 38–55 (mostly single), midbodies are in 21–23 rows and the anal scale is single. Scales are smooth to moderately rugose dorsally.

**Habits and habitat**  Unlike any other Australian elapids, apart from the other two species in its genus. It is an ambush feeder, sedentary in nature, and only appears to breed in alternate years. Prefers to seek refuge in leaf litter or loose sand, where it lies in ambush for any unsuspecting prey. Captures prey by twitching its tail rapidly when the prey is sensed. The prey is attracted to the grub-like lure, and is quickly struck by the snake, which hangs on until the prey succumbs to the fast-acting postsynaptic neurotoxins in the venom. Specimens have been found with parts of their tails missing. This suggests that, at times, the prey has bitten off part of the tail before being attacked by the snake. Part of the tail could also be lost due to incomplete skin sloughing. In any case, death adders without a tail, or part of a tail, are at a distinct disadvantage, and may eventually die from starvation. Feeds on lizards, small mammals and birds, with a greater proportion of mammals being found in larger snakes. Males mature at 24 months and females at 42 months. Up to 33 are born live in a litter. The venom is mainly neurotoxic in activity and is 1.5 times more toxic than the Indian cobra *Naja naja*.

Death adders are the closest examples of Australian snakes to the vipers or true adders. Although they are truly cobra-like or elapids, they do exhibit a small degree of fang rotation like the vipers. Scientific studies have found that this species is convergent with the vipers. They have long fangs for their size (averaging 6 mm for adult snakes). Since settlement of the Australian continent, the numbers of death adders have dramatically declined. In the south, they have been unable to adapt to habitat alteration, and now occur only in national parks and virgin scrubland with minimal alteration. The cane toad *Bufo marinus* has also been linked to a decline in their numbers in the northern parts of its range.

## Undescribed Species

There are two other undescribed species of *Acanthophis* that occur on the mainland of Australia. One is from the Barkly Tableland in Northern Territory and is intermediate in appearance between *A. praelongus* and *A. pyrrhus*.

Acanthophis (sp)
*Barkly Tableland.*

The other species Acanthophis (sp) — (Millstream area), is found in the Millstream area in Western Australia's Chichester ranges. According to David Robinson (pers com), it is found in rocky areas.

Acanthophis (sp)
*Millstream area.*

# Northern death adder

*Acanthophis praelongus* (Ramsay, 1877)
*(Highly venomous)*

**Description** A moderately stout snake, being intermediate between the common death adder and the desert death adder in rugosity of head shields, keeling of dorsal scales and numbers of ventral and subcaudal scales. Has a broad head distinct from the neck. The dorsal colouration is variable from greyish to dark reddish-brown, with darker crossbands. The head is a darker colour, being grey to black. The ventral colour is whitish, with darker spotting. Ventral scales number 122–140, subcaudals 39–57 (19–39 single), midbodies are in 23 rows (rarely 19 or 21) and the anal scale is single.

**Habits and habitat** Prefers subhumid to humid ares of the Kimberley Ranges, northern NT and possibly New Guinea. Found in grasslands, woodlands, rocky ranges and outcrops. It is an ambush feeder, lying in wait for its prey like the common death adder. From captive observations, it prefers amphibians and lizards for food, rather than mammals. Because of its liking for amphibians, it has suffered large population losses from introductions of the cane toad (*Bufo marinus*).

# Desert death adder

*Acanthophis pyrrhus* (Boulenger, 1898)
*(Highly venomous)*

**Description**  Similar appearance to the common death adder and northern death adder, but with a flatter head and more elongated body. Scales are strongly keeled dorsally, with very rugose head shields. Colour varies from a bright reddish to brown, with lighter-coloured bands whose scales can be black-tipped on the posterior side of the band. The bands are in stark contrast to the body colour when the snake is agitated and flattened out. The tail can be cream, yellow or black. Ventral colour can be cream or reddish. Ventral scales number 140–160, subcaudals 45–60 (single anterior, divided posterior), midbodies 21 and the anal scale is single.

**Habits and habitat**  Little is known about its habits because it lives in remote areas. Occurs in dry areas, favouring porcupine grass clumps, rocky outcrops, stony flats and sandy ridges. Has been found in abandoned burrows. Feeds mainly on lizards, particularly skinks and dragons. Produces litters of up to 13 liveborn young. Like the common death adder, it is an ambush feeder, lying in wait for potential food and wriggling its tail vigorously when the prey is sensed. Although an elapid, it is viper-like and is only similar in appearance and habits to the other two members of its genus. Aborigines fear it and call it *mythunda*. Its venom capabilities are assumed to be the same as the common death adder.

# Copperhead

*Austrelaps superbus* (Günther, 1858)
*(Highly venomous)*

*Lowland Copperhead*
Austrelaps superbus.

**Description** Three forms of copperheads are recognised, but they are all included under the species *superbus* as they are still formally undescribed. They are the highland form (sometimes referred to as *A. ramsayi*), the lowland form (referred to as *A. superbus*) and the pygmy form (sometimes referred to as *A. labialis*). The head is smaller and slightly distinct from the body. Attains a length of 1.7 m, but adults are normally about 0.9 m in length. The pygmy form, from the Adelaide Hills and Kangaroo Island, is much smaller, having a maximum length of about 0.7 m. The eye is large and has a round pupil. Dorsal colours are from black, brown, tan, coppery to light grey, sometimes having a dark vertebral stripe and dark band across the nape, especially in juveniles. The ventral surface is cream to grey, often with a red colour at the edges. The labial scales give a striking barred appearance. Ventral scales number 140–165, subcaudals 35–55 (single), midbodies 15 (rarely 13 or 17) rows and the anal scale is single.

**Habits and habitat** Has the lowest tolerance for cool temperatures of all the large Australian elapids. Active earlier in spring and later in autumn than most other snakes and has even been seen basking in a sunny spot out of the wind in winter. Becomes nocturnal in summer. Inoffensive, shy and retiring. Occurs in large colonies in woodlands, tussock grasslands, heaths, highlands and moist low-lying areas. Produces up to 20 live young in a litter. Attains sexual maturity at 2 years and mates in spring, when males can occasionally be seen in combat (not observed for pygmy form yet). A skewed sex ratio appears to exist in favour of males. It is a forager and feeds mainly on small lizards, frogs and tadpoles (most prey items are small) and is closely associated with moisture or water. The pygmy form appears to feed more frequently than the other two forms. Capable of flattening the whole body as a deterrent when threatened. The venom is strongly neurotoxic, weakly coagulant, strongly anticoagulant, blood destroying and muscle destroying, and is about the same toxicity as the Indian cobra *Naja naja*.

*Pygmy Copperhead*
Austrelaps (sp).

# White-crowned snake

*Cacophis harriettae* (Krefft, 1869)
*(Venomous, not dangerous)*

**Description**  A small snake not exceeding 0.4 m in length. Dorsal colouration is dark brown or steely grey, with a white or creamish band around the head, forming a complete broad collar on the nape at least 4 scales wide. The top of the head is glossy black. Some body scales possess light longitudinal centres, which form narrow bands, especially anteriorly. The ventral surface is grey. The scales are smooth and the ventrals number 170–200, midbodies 15 rows, sub-caudals 25–45 (divided) and the anal scale is divided.

**Habits and habitat**  Nocturnal, preferring wet subtropical to tropical environments. An egg-laying species producing 2–10 eggs, with an average of 5.1, mainly in autumn. Females grow much larger than males. Males mature at 20 months and most females at 32 months. Feeds mainly on diurnal lizards (*Lampropholis* and *Leiolopisma*) by foraging, but has been recorded feeding on blind snakes, lizard eggs and frogs. Feeds all months of the year, probably due to the combination of the mild climate throughout its range and its small size, allowing it to heat more rapidly than large snakes. When provoked, it raises its head and at the same time points its head down to display the light-coloured crown on its head, then strikes repeatedly, usually without opening its mouth.

# Dwarf-crowned snake

*Cacophis krefftii* (Günther, 1863)
*(Venomous, not dangerous)*

**Description**  A small snake which grows up to 0.33 m. Dark grey to black dorsally. A dark median line occurs beneath the tail. Ventral scales are white medially and edged with black, forming a sawtooth pattern. A narrow pale white band 0.5–2 scales wide stretches across the nape, extending forward on the side of the head, where it is broken by black scales, then on to the snout. Midbody scales are smooth and number 15, ventrals 140–160, subcaudals 25–40 (divided) and the anal scale is divided.

**Habits and habitat**  Inhabits a narrow coastal strip along the eastern part of the mainland, where it prefers a wet subtropical climate. Found under rotting logs, bark or leaf litter. Mainly nocturnal and secretive in nature. It is oviparous, with an average clutch size of 3.2, which occurs mainly in autumn. Females grow much larger than males. Both males and females mature at 20 months. Feeds mainly on diurnal lizards (*Lampropholis* and *Leiolopisma*) by foraging, but has been recorded feeding on blind snakes, lizard eggs and frogs. Feeds all months of the year, probably due to the combination of the mild climate throughout its range and its small size, allowing it to heat more rapidly than large snakes. Its defence posture is similar to that of the white-crowned snake.

# Golden-crowned snake

*Cacophis squamulosus* (Duméril, Bibron and Duméril, 1854)
*(Venomous, not dangerous)*

**Description** Grows to 0.75 m in length and averages 0.5 m. Has a slender body with an angular head distinct from body. Dark brown to grey dorsally, with a brown to yellow stripe extending from the snout to the nape on either side of the head, but not joining to form a complete collar. The lips are barred with dark brown. The ventral surface is pink to red with black patches and spots concentrated about the midline. A black stripe occurs under the tail at the junction of the divided subcaudal scales. The eyes are small with elliptical pupils. Midbody scales are in 15 rows, ventrals number 165–185, subcaudals 30–50 (divided), and the anal scale is divided.

**Habits and habitat** Shelters under logs, stones or leaf litter during the day and is secretive and nocturnal when active. Prefers well-watered areas. Like other members of this genus, it has a fierce-looking threat pose, but rarely attempts to bite. It is oviparous, producing between 2 and 15 eggs (average 6.2), mainly in autumn. Females grow much larger than males. Males mature at 20 months whilst females mature at 32 months. Feeds mainly on diurnal lizards (*Lamprophalis* and *Leiolopisma*) by foraging, but has been recorded feeding on blind snakes, lizard eggs and frogs. Feeds all months of the year, probably due to the combination of the mild climate throughout its range and its small size, allowing it to heat more rapidly than large snakes. Uses constriction to restrain prey whilst the venom immobilises it.

# Small-eyed snake

*Cryptophis nigrescens* (Günther, 1862)
*(Venomous)*

**Description** Relative eye size is comparable to other elapids. Grows to 1.2 m in length and averages about 0.5 m. Has a robust body with a head distinct from the neck. Its colour is a shiny blue-black dorsally and white, cream or pink ventrally, sometimes with darker blotches. Midbody scales are in 15 rows, ventrals 165–210, subcaudals 30–46 and the anal scale is single. The nasal scale contacts the preocular scale.

**Habits and habitat** Prefers sandstone, well-timbered and rocky areas. Mainly nocturnal, but found during the day under rocks, in crevices, earth cracks or under the bark of fallen trees. They have been found hibernating together in large numbers. Bears up to 5 live young. Adult males are larger than females. Male combat has been observed in this species. Feeds on lizards (89 per cent), occasionally snakes (5 per cent — *Ramphotyphlops*, *Drysdalia*, *Furina*, *Notechis* and *Vermicella* spp. all being recorded) and very rarely frogs. Its venom is only about 21 per cent as toxic as the Indian cobra *Naja naja*. One death has resulted from a bite by this snake.

# Northern small-eyed snake

*Cryptophis pallidiceps* (Günther, 1858)
*(Venomous, not dangerous)*

**Description** Relative eye size is comparable to other elapids. Grows to 0.63 m in length. Head is slightly distinct from the neck, with a moderately slender body. A glossy black or grey-black dorsal colour, lower lateral scales tinged with orange and often has a paler head. The upper labials are whitish. The ventral scales can be cream, grey or pink, often bearing dark flecks posteriorly, extending to the midline of the subcaudal scale. Midbody scales are in 15 rows, ventrals number 160–180, subcaudals 45–60 (single) and the anal scale is single.

**Habits and habitat** Shelters beneath rocks, in termite mounds, logs and other debris. Predominantly nocturnal. It is live bearing and has an average litter size of 4. Adult females are larger than males. Feeds mainly on lizards (89 per cent), occasionally snakes (5 per cent — *Ramphotyphlops*, *Drysdalia*, *Furina*, *Notechis* and *Vermicella* spp. all being recorded) and very rarely frogs.

# Black whip snake

*Demansia atra* (Macleay, 1884)
*(Venomous, potentially dangerous)*

**Description** The head is long and narrow, distinct from neck and body, which is long and slender. The tail is thin and whip-like. Grows to 1.8 m in length. Dorsal colour is dark brown, reddish-brown or black, becoming reddish posteriorly. Individual scales may possess posterior black margins. The head is a coppery colour. The ventral colour is yellowish-grey to greenish-grey, with anterior scales edged in black. The underside of the tail is reddish. The side of the face has a trace of contrasting dark and light markings. Midbody scales are in 15 rows, ventrals number 160–220, subcaudals 70–95 (divided) and the anal scale is divided.

**Habits and habitat** Prefers drier subhumid habitats throughout its range. Diurnal, becoming nocturnal on hotter nights. It is oviparous and lays from 4–20 eggs, breeding throughout the year. Exhibits male combat. Feeds mainly on lizards (73 per cent), but has been recorded taking small frogs (27 per cent). Males attain larger body size than females.

# Marble-headed whip snake

*Demansia olivacea* (Gray, 1842)
*(Venomous)*

Demansia olivacea
olivacea.

**Description** Head is barely distinct from the neck and long thin body. Grows to 0.85 m in length. Dorsal colour is olive-grey, brownish or reddish, often with dark colouration at the base of each scale. The snout and lips are a mottled blackish-brown. Head often possesses blackish marbling or blotching from the snout to the parietal scales. The side of the head is marked with a purple-brown streak, extending from the eye to the angle of the mouth, bordered above by a narrow white streak and below by a paler colour, diffusing into the mottling around the lips. There are two grey streaks on the chin, and the throat has blackish-brown spots. The ventral surface of the body is creamish or white, and the tail is yellow-green or reddish. Midbody scales are in 15 rows, ventrals number 160–210, subcaudals 65–105 (divided) and the anal scale is divided.

**Habits and habitat** A slender, fast-moving snake. Occurs in semi-arid to tropical areas in woodlands, stony ranges and savannah grasslands. Mainly diurnal, becoming nocturnal on warm nights. It is egg laying, with a clutch size of 3–4 eggs, and appears to reproduce throughout the year. Feeds mainly on small lizards (86 per cent). Males attain a larger body size than females.

### Subspecies

***Demansia olivacea olivacea*** (Gray, 1842).
Brownish dorsally. Occurs throughout the Kimberleys, top end of Northern Territory and north-western Qld.

***Demansia olivacea calodera*** (Storr, 1978).
Olive-grey dorsally, with a pale-edged nuchal bar. Occurs in coastal

regions and hinterland of WA from North-West Cape to Shark Bay.

Demansia olivacea 'calodera.

### *Demansia olivacea rufescens* (Storr, 1978).

Reddish-brown to coppery-brown dorsally. Found in the Pilbara region of WA.

Demansia olivacea rufescens.

# Papuan whip snake

*Demansia papuensis* (Macleay, 1877)
*(Venomous)*

Demansia papuensis
papuensis.

**Description** A large, slender, nervous and fast-moving snake, growing to 1.8 m in length. The front of the head is paler than the rest of the body.

**Habits and habitat** Occurs in dry to moist areas of its range.

### Subspecies

***Demansia papuensis melaena*** (Storr, 1978).
Dark brown to blackish dorsally, becoming redder posteriorly. The eye may be narrowly margined with cream or white. The ventral surface is grey, becoming paler towards the head. The ventral surface beneath the tail is reddish. Ventrals number 192–220, sub-

Demansia papuensis
melaena.

caudals 78–105, the anal scale is divided and midbodies are in 15 rows.

### *Demansia papuensis papuensis* (Macleay, 1877).

Paler than *Demansia papuensis melaena*. Ventrals number greater than 220, subcaudals 78–105, the anal scale is divided and midbodies are in 15 rows. Extends into southern New Guinea.

# Yellow-faced whip snake

*Demansia psammophis* (Schlegel, 1873)
*(Venomous)*

Demansia psammophis
psammophis.

**Description** Consists of three subspecies. A fast-moving snake, growing to about 1.0 m in length. Highly variable in colour, ranging from a dorsal colour of light steely-grey to grey-green. Each scale is dark edged and the head is reddish. The tail is a lighter brown than the body. A yellow-edged dark bar extends from nostril to nostril around the front of the snout. A yellow-edged dark streak curves back to the angle of the mouth from the eye, like a comma. The eye is more or less surrounded by a yellow colour.

**Habits and habitat** Found in a wide range of habitats from coastal to the arid interior. It is diurnal. Males attain a larger body size than females. Sexually mature at about 20 months for both sexes. It is egg laying, producing from 3–20 eggs. Has been recorded as a communal egg layer. Its mating period is in spring. Feeds mainly on lizards (90 per cent).

### Subspecies

#### *Demansia psammophis psammophis*
Basic colour pale grey, bluish-grey, olive to brownish-grey. A brown tinge extends from behind the neck to about half the body length. Dark skin between the scales is visible, forming a network pattern. Ventral surface is greyish, tending yellowish towards the tail. Juveniles have been observed to use constriction to restrain prey whilst the venom immobilises it.

#### *Demansia psammophis cupreiceps* (Storr, 1978).
Dorsally, the body is a bluey-greenish colour, becoming mustard-

brown posteriorly. The dark skin between the scales is more pronounced than in *D. p. psammophis.*

Demansia psammophis cupreiceps.

**Demansia psammophis reticulata** (Gray, 1842).
Pale or dark olive-brown to olive. The scale pattern is similar to *D. p. cupreiceps.*

Demansia psammophis reticulata.

## *Demansia simplex* (Storr, 1978)
### *(Venomous)*

**Description**  More robust in appearance than other members of the genus. Grows to 0.6 m in length. The head is barely distinct from the body. Its colour is greyish-brown dorsally to grey on the sides. It has a contrasting white ventral surface and a pale-edged, dark comma from the eye to the angle of the mouth. Midbody scales are in 15 rows, ventrals number 140–150, subcaudals 50–65 and the anal scale is divided.

**Habits and habitat**  Normally diurnal. Feeds on small lizards. Can be active on warm nights. It is an egg layer.

# Collared whip snake

*Demansia torquata* (Günther, 1862)
*(Venomous)*

**Description** Head is barely distinct from the body. Grows to
0.85 m in length. Grey-brown above, with the top of the head
slightly darker and variably bordered by a series of black and yellow
markings, which become less distinct in older specimens. A light-
edged, narrow dark line extends from nostril to nostril across the
snout. A wide dark streak bordered by yellow extends from the eye
to the angle of the mouth in the shape of a comma. The ventral
surface is grey, being darker in the middle. Midbody scales are in 15
rows, ventrals number 185–220, subcaudals 70–90 and the anal
scale is divided.

**Habits and habitat** A fast-moving snake, found in a variety of
habitats (subhumid to arid), including grassy woodlands and grasslands. It is diurnal and feeds on lizards (100 per cent). Mating occurs
from September to November and it lays eggs, producing 2–8 in a
clutch. Males attain a larger body size than females.

# De Vis's banded snake

*Denisonia devisii* (Waite & Longman, 1920)
*(Venomous)*

**Description**  A broad depressed dark brown head, distinct from a robust body. Total length is about 0.5 m. Light brown dorsally, with darker crossbands. The scales in the darker crossbands sometimes have lighter centres. The darker crossbands are often interrupted towards the centre anteriorly. The labial scales are distinctly striped. Ventral surface is creamish. Midbody scales are in 17 rows, ventrals number 120–150, subcaudals 20–40 (single) and the anal scale is single.

**Habits and habitat**  A nocturnal species, found under logs, debris or in earth cracks. It is live bearing, having 3–9 in a litter. Feeds mainly on frogs, with the occasional lizard. If agitated, it will flatten its entire body and will strike anything within reach. Adult males and females are the same size.

# Rosén's snake

*Denisonia fasciata* (Rosén, 1905)
*(Venomous)*

**Description** The head is depressed and distinct from a robust body. Grows to about 0.6 m in length. Ground colour is light brown. The head has a number of scattered darker brown spots and a dark brown streak extending from the nose through the eyes to the neck. The remainder of the body has scattered darker brown spots. The ventral surface is white or cream. Midbody scales are in 17 (rarely 19) rows, ventrals number 140–185, subcaudals 20–40 (single) and the anal scale is single.

**Habits and habitat** Nocturnal and feeds on lizards (especially agamids). Nothing is known about its reproductive mode. Adult males are the same size as females.

# Ornamental snake

*Denisonia maculata* (Steindachner, 1867)
*(Venomous)*

**Description** A broad head distinct from a short robust body. Grows to about 0.5 m. Light brown to dark brown, sometimes grey-brown dorsally, with lighter lateral colours, which can be flecked with darker streaks or spots. The crown of the head is a darker brown or black and the side speckled with brown and cream. The labials are distinctly barred. The ventral surface is white or cream. Midbody scales are in 17 rows, ventrals number 120–150, sub-caudals 20–40 (single) and the anal scale is single.

**Habits and habitat** Nocturnal species, feeding mainly on frogs and occasionally lizards. Shelters under litter and fallen timber during the day. It is live bearing, producing 3–11 young in a litter. When agitated, it flattens its whole body and strikes without warning. Its venom could produce serious effects. Adult males are the same size as females.

# Little spotted snake

*Denisonia (Rhinoplocephalus) punctata* (Boulenger, 1896)
*(Venomous)*

**Description** Head is depressed and slightly distinct from the neck. Grows to 0.5 m. Dorsally, reddish-brown except for the upper labials, which are white or pale cream. There is a short dark brown or black streak from the rostral through to the eye. A few irregular black spots on the dorsal and lateral surfaces on the head and neck. The midbody scales are in 15 rows, ventrals number 146–183, sub-caudals 20–42 (single) and the anal scale is single.

**Habits and habitat** Prefers drier parts of its range. It is nocturnal and feeds mainly on lizards (especially agamids). Has been recorded feeding on blind snakes. Is live bearing, producing 2–5 in a litter.

# Crowned snake

*Drysdalia coronata* (Schlegel, 1837)
*(Venomous)*

**Description** Head is slightly distinct from a moderately slender body. Grows to 0.6 m. Yellowish-brown, grey or olive-brown to dark brown dorsally. The tail is often a lighter yellowish or reddish colour. The head is a darker grey, completely enclosed by a black surround, which appears as a thin line on the rostral scale and along the side of the head through the eye to an expanded bar on the nape. The labials are white or yellowish. Midbody scales are in 15 rows, ventrals number 130–160, subcaudals 35–55 (single) and the anal scale is single.

**Habits and habitat** Mainly nocturnal, feeding on small skinks (50 per cent) and frogs (50 per cent). Breeds each year, being live bearing and producing 3–9 in a litter during late summer. The ratio of males to females born is 1:1. Females are the same size as males when mature. Sexual maturity takes about 1.5 years.

# White-lipped snake

*Drysdalia coronoides* (Günther, 1858)
*(Venomous)*

**Description** Head is slightly distinct from a moderately slender body. Grows to 0.5 m in length. Both its dorsal and ventral colours are highly variable. Dorsally, colour ranges from light grey, brown, russet to black. The ventral surface may be creamish, yellowish or pinkish. A thin, either continuous or broken white line, bordered above by black, extends from the snout along the upper labials to the neck. Midbody scales are in 15 rows, ventrals number 120–160, subcaudals 35–70 (single) and the anal scale is single.

**Habits and habitat** Mainly nocturnal and feeds on small skinks (90 per cent), their eggs or frogs. During the day it seeks refuge under rocks, logs and litter. Breeds each year on the mainland, but in Tasmania only reproduces once every 2–3 years. It is live bearing, producing 2–10 young in a litter in late summer. The ratio of males to females born is 1:1. Sexual maturity is reached after 2.5 years. The size of adult females is about the same as adult males.

# Masters' snake

*Drysdalia mastersi* (Krefft, 1866)
*(Venomous)*

**Description** Head is slightly distinct from a moderately slender body. Grows to 0.4 m in length. Dorsal colour is grey-brown, sometimes with a red sheen. The head is dark, bordered on the side by a black-edged stripe extending from the rostral, under the eye to the neck, and on top by a light yellow, orange or brown collar on the nape. The ventral surface is yellow, with darker flecks along the outer edges of the ventral scales. Midbody scales are in 15 rows, ventrals number 130–160, subcaudals 32–55 (single) and the anal scale is single.

**Habits and habitat** Occurs on coastal dunes, limestone outcrops and sand plains supporting heathlands and mallee. It is nocturnal and feeds mainly on small lizards (90 per cent). Breeds each year and is live bearing, producing 2–3 young in a litter in late summer. The ratio of males to females born is 1:1. At maturity, males and females are the same size.

*Drysdalia rhodogaster* (Jan, 1863)
*(Venomous)*

**Description** Head is distinct from a robust body. Grows to
0.45 m in length. Dorsal colour is olive grey to brownish. Head is
dark with a mottling towards the snout and a sharply defined pale
yellow band 2–3 scales wide on the nape. A dark brown streak runs
from the snout through the nostril and eye. Midbody scales are in 15
rows, ventrals number 140–160, subcaudals 40–55 (single) and the
anal scale is single.

**Habits and habitat** Shelters under rocks, logs and other debris.
Prefers dry habitats of woodlands and heathlands throughout its
range. Mainly diurnal, but becomes nocturnal in warmer weather.
Feeds mainly on lizards (90 per cent). Breeds each year and is live
bearing, producing up to 6 young in a litter in late summer. At
maturity, males and females are the same size.

## *Echiopsis atriceps* (Storr, 1980)
### *(Venomous)*

**Description**  Head is distinct from a robust body. Grows to about 0.5 m. Dorsally, brown to dark brown, with a light reddish-brown ventral surface. The head is black on the top and sides, with a narrow white edge on the upper lips. Midbody scales are in 19 rows, ventrals number 175–180, subcaudals 45–50 (single) and the anal scale is single.

**Habits and habitat**  Little is known about this snake as only a few specimens have been recorded. Its bite has been reported as producing severe symptoms.

# Bardick

*Echiopsis curta* (Schlegel, 1837)
*(Venomous)*

**Description** Head is depressed and distinct from a stout body. Grows to 0.62 m. Dorsal colour is variable, ranging from grey, olive-brown to red. The labials are usually dotted with white. The ventral surface is grey to brown, with each ventral scale edged with brown posteriorly. The subcaudals are tending to yellow. Midbody scales are in 19 rows, ventrals number 130–145, subcaudals 30–40 (single) and the anal scale is single.

**Habits and habitat** Seeks refuge under logs and other ground debris, and occurs in the more arid parts of its range. It is nocturnal and feeds on lizards (52 per cent), frogs (31 per cent), small mammals (13 per cent), and occasionally birds and insects. Is live bearing and produces litters of 3–14 young. Newly born snakes grow to two-thirds of average adult size in the first year. Males mature in their second year and females in their third year. Adult females are the same size as males. Since its venom has some similarities to *Acanthophis* venom, bites may require medical treatment.

# Little brown snake

*Elapognathus minor* (Günther, 1863)
*(Venomous)*

**Description**  Head is slightly distinct from a robust body. Grows to 0.45 m. Dorsal colour is a uniform brown, tending to greenish on the lateral surface, with a bright red on the tail. The supralabials are yellowish. The nape often has a yellow-black collar. The ventral scales are cream to green, with a black crescent-shaped area anteriorly, forming a series of black bars medially, which become fainter on the tail. Midbody scales are in 15 rows, ventrals number 120–130, subcaudals 50–70 (single) and the anal scale is single.

**Habits and habitat**  Shelters in low, dense vegetation. Very little is known about this snake. Confined to the lower south-east of WA. It is a live bearer and produces 8–12 young, with oviposition occurring in mid-summer. Diurnal, and feeds on skinks and frogs. At maturity, males are the same size as females.

# Red-naped snake

*Furina diadema* (Schlegel, 1837)
*(Venomous)*

**Description** Head is slightly distinct from a slender body. Grows to 0.45 m. Dorsally, rich reddish-brown, with a black skin showing between the scales. The head and nape are shiny black, separated by a crescent of orange. The ventral surface is white or cream. Mid-body scales are in 15 rows, ventrals number 160–210, subcaudals 35–70 (divided) and the anal scale is divided.

**Habits and habitat** Nocturnal, feeding all year round on small skinks (especially *Lampropholis* spp.). It is oviparous, producing 1–5 eggs in a clutch each season. Sexual maturity is attained in the second year after hatching. Found under rocks, logs, in earth crevices and other ground debris. Often found in association with termite and ant colonies. When alarmed, it raises the forepart of its body.

# Orange-naped or moon snake

*Furina ornata* (Gray, 1842)
*(Venomous)*

**Description**  Head is slightly distinct from a slender body. Grows to 0.7 m in length. Orange or reddish-brown dorsally. The head is dark brown or black, tending to be lighter towards the snout. The lips are creamish. A strong yellow-orange band divides the black-coloured head and nape. The ventral surface is cream. Midbody scales are in 15 or 17 rows, ventrals number 160–240, subcaudals 35–70 (divided) and the anal scale is divided.

**Habits and habitat**  Occurs in subhumid to arid areas throughout its range. A nocturnal species which shelters beneath rocks and other debris during the day. Feeds on small skins (especially *Lampropholis* spp.). It is oviparous, producing 3–6 eggs. Attains sexual maturity in its second year. It is also thought that females may be capable of producing two clutches per year.

# Yellow-naped snake

*Glyphodon barnardi* (Kinghorn, 1939)
*(Venomous)*

**Description**  Head is barely distinct from a slender body. Grows to 0.5 m. Dorsally, brownish to black, the scales being light edged. The head is blackish, with a lighter yellow or brown collar on the back of the head to the neck. The collar is less distinct in older specimens. The ventral surface is whitish or cream. Midbody scales are in 15 rows, ventrals number 170–200, subcaudals 35–50 (divided) and the anal scale is divided.

**Habits and habitat**  Nocturnal, sheltering under logs during the day. It is egg laying, producing from 6–10 eggs. Feeds mainly on skinks, especially *Sphenomorphus* spp.

# Dunmall's snake

*Glyphodon dunmalli* (Worrell, 1955)
*(Venomous)*

**Description** Head is distinct from a moderately robust body. Grows to 0.75 m in length. A uniform dark grey-brown dorsally, with a white ventral surface. Midbody scales are in 21 rows, ventrals number 175–190, subcaudals 35–50 (divided) and the anal scale is divided.

**Habits and habitat** Occurs in dry brigalow scrub. Its habitat is becoming threatened. Little is known about this nocturnal species. Feeds on skinks, especially *Sphenomorphus* spp. The black tree skink *Egernia striolata* has also been recorded as a prey item. It is oviparous and produces 6–10 eggs in a clutch.

# Brown-headed snake

*Glyphodon tristis* (Günther, 1858)
*(Venomous)*

**Description** Head is indistinct from a slender body. Average length is about 0.7 m, but it can grow to 1.0 m. Dorsal colour is blackish-brown, with a white trailing edge on each scale, which is even more conspicuous on the sides. A light brownish or yellow collar occurs on the nape. The ventrals are white or cream, with a brown patch on the outer edges. The margins of the subcaudal scales are strongly marked with brown. Some dark markings around the chin. The lower labials and side of neck are creamish. Midbody scales are in 17 rows, ventrals number 160–190, subcaudals 30–60 (divided) and the anal scale is divided.

**Habits and habitat** Occurs in woodlands, forests and vine thickets. Nocturnal and feeds on skinks, especially *Sphenomorphus* genus. It is oviparous, producing 6–10 eggs in a clutch. Extremely nervous when threatened and thrashes about wildly when provoked.

# Grey snake

*Hemiaspis damelii* (Günther, 1876)
*(Venomous)*

**Description** Head is slightly distinct from a moderately robust body. Grows to 0.75 m in length. Dorsally, grey to olive-grey, sometimes with a black spot at the base of each scale. The head is black in juveniles, tending to fade into a dark nuchal band with age, in some cases disappearing altogether. The ventral surface is white or cream, sometimes with darker flecks on the chin. The subcaudals are strongly edged with brown. Midbody scales are in 17 rows, ventrals number 140–170, subcaudals 35–50 (single) and the anal scale is divided.

**Habits and habitat** Either crepuscular or nocturnal and sometimes diurnal. Seeks refuge in woodlands. It is a live bearer and produces 6–12 young. Reproduces each year. Feeds on frogs and occasionally small lizards. Adult females are larger than males.

# Marsh or black-bellied swamp snake

*Hemiaspis signata* (Jan, 1859)
*(Venomous)*

**Description** Head is slightly distinct from a moderately robust body. Grows to 0.9 m in length, although the average is around 0.5 m. Dorsally, olive-brown to grey, with a darker head. Two white or yellow streaks on the side of the head, one from the eye to the neck and the other on the supralabials. Some of the NSW population may tend toward melanism. The ventral surface is black or grey. In Queensland, the dorsal colour may be russet or pink and the ventral surface cream or salmon. Midbody scales are in 17 rows, ventrals number 150–170, subcaudals 40–60 (single) and the anal scale is divided.

**Habits and habitat** Either crepuscular, diurnal or nocturnal in warmer weather. Prefers swampy marshes, where it may congregate in small colonies, rocky ridges or sandy dunes. Its reproduction is geared to a seasonal pattern and it is a live bearer, producing 4–20 young. Females are gravid between November and March. Feeds on small lizards (especially *Lampropholis* spp.) and frogs, and has been recorded feeding on other small snakes. Male combat has been observed in this species. Adult males are larger than females.

# Pale-headed snake

*Hoplocephalus bitorquatus* (Jan, 1859)
*(Venomous)*

**Description** Broad head is distinct from a moderately robust body. Grows to 1.2 m. Dorsally, either brown or grey. The head is grey, with a white or cream band on the nape, bordered posteriorly with a dark band. There are numerous black spots on the head, especially adjacent to the lighter band on the nape. The ventral surface is creamy grey, sometimes with darker flecks. Midbody scales are in 19 or 21 rows, ventrals number 190–225, subcaudals 40–65 (single) and the anal scale is single.

**Habits and habitat** Nocturnal and arboreal. Found in wet sclerophyll and dry eucalypt forests, where it shelters under loose bark or in tree hollows, especially in the vicinity of water. It is a live bearer and produces 2–11 young, usually in late summer. The young mature in 3 to 4 years. Females reproduce only every 2 years or less. Adult females are larger than males. Feeds mainly on arboreal frogs; small lizards and mammals may be taken. It has a painful bite, and may cause serious illness.

# Broad-headed snake

*Hoplocephalus bungaroides* (Schlegel, 1837)
*(Highly venomous)*

**Description** Broad head is distinct from a moderately robust body. Grows to 1.0 m in length. Dorsally, it is jet black, with numerous yellow spots, which form a series of irregular thin crossbands. The ventral surface is grey or grey-black, sometimes with yellow blotches. Midbody scales are in 21 rows, ventrals number 200–230 (keeled), subcaudals 40–65 (single) and the anal scale is single.

**Habits and habitat** A nocturnal species which seeks refuge under sandstone ridges. The main population occurs in Hawkesbury sandstone formations. The whole population occurs within a 250 km radius of Sydney. It is live bearing, producing between 2 and 11 young per litter. Adult females breed only in alternate years or greater, and are larger than males. Feeds mainly on lizards and occasionally frogs. Particularly nervous and excitable. If provoked, may stand its ground and offer a threat pose, with its body arched in an S-shape. Will readily strike at anything considered to pose a threat. Its bite may cause acute symptoms and serious illness. Although the venom is not normally regarded as potentially fatal, in some circumstances it may be fatal. It can be easily confused with the harmless diamond python *Morelia spilotes spilotes*. Its venom is powerfully coagulant, neurotoxic and weakly blood destroying.

Classified as a threatened species, based on its close proximity to urban development, its small range and the continued degradation of its habitat by landscapers who use weathered sandstone for bushrock artifacts in their gardens. Even the habitats 'protected' in national parks in its range are being illegally vandalised. The fact that it is a venomous snake also makes it difficult to value in the eyes of the wider community. Fortunately, it has been bred a number of times in captivity, and if sufficient importance is put into this practice, its continued existence will be ensured, albeit in an alien environment, while the borrowed time is used to conserve existing habitats and rehabilitate others. Cooperation between captive breeders and professional researchers is an important part of the conservation strategy to save this species. It also involves the understanding of the relevant statutory body in the complete requirements for its survival and dedication of funds for its conservation.

# Stephens's banded snake

*Hoplocephalus stephensi* (Krefft, 1869)
*(Highly venomous)*

**Description** Broad head is distinct from a moderately robust body. Grows to 1.2 m in length. Dorsally, dark brown to bluey-black, with a series of narrow light brown to orange crossbands, which become broken and less distinct posteriorly. Some specimens lack bands altogether. There is usually a light brown blotch on top of the head, which is otherwise black with brown-cream markings on either side of the nape. The lips are barred with black and cream. The ventral surface is white or cream with black blotches. Midbody scales are in 21 rows, ventrals number 220–250, subcaudals 50–70 (single) and the anal scale is single.

**Habits and habitat** Mainly nocturnal and mostly arboreal. Occurs in rainforests, wet or dry sclerophyll forests and rocky outcrops. Shelters beneath loose bark, hollow logs or rocks. Occasionally found basking during the day in mild weather. Mating occurs in late spring to early summer. It is a live bearer and produces 3–8 young (average of 6) in February and March. Adult females breed only in alternate years or longer. Adult females are larger than males. Feeds on lizards, small mammals, frogs and occasionally birds. The venom contains powerful procoagulant, moderate neurotoxic, strong haemorrhagic and weakly haemolytic activity, and therefore should be considered dangerous. It is about 36 per cent as toxic as the venom of the Indian cobra *Naja naja*.

# Western black-naped or long-nosed snake

*Neelaps bimaculatus* (Duméril, Bibron and Duméril, 1854)
*(Venomous)*

**Description** Head is indistinct from a slender body. Grows to 0.45 m in length. Dorsally, reddish-brown, orange to pinkish, with each scale edged in dark reddish-brown. Has a black blotch on the head, starting at the front edge of the frontal scale and extending to the rear edge of the parietals, with a black nuchal band. The ventral surface is white or cream. Midbody scales are in 15 rows, ventrals number 175–235, subcaudals 15–35 (divided) and the anal scale is divided. Males tend to have more subcaudals and fewer ventrals than females.

**Habits and habitat** A small burrowing snake, occurring in sub-humid to semi-arid zones throughout its range. Prefers sandy soils with low, open vegetation. It is egg laying and produces 2–6 eggs. Feeds on lizards. Adult males are smaller than females.

# Western black-striped snake

*Neelaps calonotus* (Duméril, Bibron and Duméril, 1854)
*(Venomous)*

**Description** Head is indistinct from a moderately robust body. Australia's smallest snake, attaining only 0.28 m in length. Dorsal colour is creamish, with each scale having reddish-orange edges. The nose is tipped with black. A black half-band encloses the eyes and top of the head, spanning from the front edge of the frontal scale to the rear edge of the parietals, and a black band is on the nape about 3 scales wide. A vertebral stripe of black scales with white centres extends from the nape to the tail tip. The ventral surface is creamish. Midbody scales are in 15 rows, ventrals number 120–150, subcaudals 20–40 (divided) and the anal scale is divided.

**Habits and habitat** A small burrowing snake, occurring in coastal dunes, sand plains, heathlands and eucalypt woodlands. Because of its small range and proximity to human development, it is potentially endangered. It is an egg layer and produces 2–5 eggs. Feeds on small lizards, especially *Lerista* sp. Adult males are smaller than females.

# Krefft's tiger snake

*Notechis ater ater* (Krefft, 1866)
*(Highly venomous)*

**Description** Broad flat head is distinct from a robust body. Grows to 1.02 m in length. Dorsally, jet black or black, with creamish bands over 75 per cent of its body. The tail is black. Many specimens have white or cream markings around the chin and infralabials. Juveniles have creamy white bands. The ventral surface is either black or grey, tending to creamish before sloughing. Midbody scales are in 17 rows, ventrals number 163–173, subcaudals 41–50 (single) and the anal scale is single (occasionally divided).

**Habits and habitat** Diurnal species, occurring in a confined area of the lower Flinders Ranges where the rainfall averages 600 mm per year. More recent studies have observed it along creeklines draining from high rainfall areas. A separated population occurs near the mouth of the Broughton River, although this population is connected by the same drainage systems. The main population occurs along creeks in the high rainfall area (above 600 mm). These areas typically have *E. camaldulensis* and *E. cladocalyx* along the creeks and *E. goniocalyx* on the hill slopes. An account by Robert Bruce from Arkaba Station in 1902 tells of black snakes occurring in gum-lined creeks, some distance from the present known population. There is record of a snakebite from the northern Flinders Ranges that tested positive for tiger snake. There is also a single unconfirmed report from Wilpena Pound of a snake resembling a tiger snake.

Found in debris in creekbeds, rocky screes on the slopes and tree and shrub growth along the creeks out on the plains. Restricted to watercourses fed by natural springs. Feeds on frogs, tadpoles, small mammals and occasional birds, including the black duck *Anas super-*

*cilosa*. Found in the water after November when the water temperature rises above 17°C. At that time of the year, it continually shuttles in and out of the water searching for food, and then warming up again after cooling in the water. Its black colour gives it a distinct advantage in enabling it to quickly re-elevate its body temperature. It is a live bearer and produces 8 to 15 young.

Although it is represented in Mt Remarkable National Park, this species is under threat because its range is so small and bushfires, which occasionally raze this park, have the potential of further reducing their numbers. To ensure the species' conservation, the cooperation of the landowners in its range is necessary. This would involve fencing creeks to halt erosion and degradation, and allow the restoration of natural vegetation, which is currently being overgrazed. Its venom is the most toxic of all the tiger snakes.

# Peninsula tiger snake

*Notechis ater niger* (Kinghorn, 1921)
*(Highly venomous)*

**Description** Blunt head is distinct from a robust body. Averages
1.1 m in length. Roxby Island specimens are much smaller, averaging 0.86 m in length. Dorsally, generally jet black, sometimes with
white or cream markings around the lips and chin. On Kangaroo
Island, specimens are highly variable in colour, often exhibiting
banding and uniform brown colours. The ventral surface is dark
grey to black, with some specimens on Kangaroo Island even possessing red bellies. The ventral surface becomes much lighter prior
to shedding. Juveniles nearly always have banding. Midbody scales
are in 17, 18, 19 and rarely 21 rows, ventrals number 160–184,
subcaudals 45–54 (single) and the anal scale is single.

**Habits and habitat** Occurs in coastal dunes and associated
heath, coastal mallee, eucalypt forests, open grasslands and limestone outcrops. Most of the habitats experience between 600 and
800 mm of rainfall per year; however, the small island populations
survive in much drier conditions. The island populations are more
specialised in prey items. For instance, on Roxby Island, lizards
comprise most of the diet. On Reevesby Island, migratory storm
petrels are the main diet of adult snakes. On Franklin and Hopkins
Islands, mutton birds are the main diet of adult snakes. Will take
other prey items, such as small skinks, small birds, small mammals
(introduced *Mus musculous*) and even larger mammals such as the
grey rat, sticknest rat and bandicoot. On the mainland and Kangaroo Island, the range of prey items is greater and the predators are
more varied. Has been observed swimming in the sea, kilometres
from shore. This suggests that the species could travel from island to
island and the mainland. However, size difference of specimens

from each island contradicts this suggestion. Males are larger than females in both length and weight. The venom of the Reevesby Island tiger snake is more toxic than *N. scutatus* venom. The freeze-dried venoms of all *N. ater niger* populations is white in colour, except for the Kangaroo Island population, which is a cream colour similar to that of *N. scutatus*. It is live bearing, producing up to 20 young in a litter. The venom of the Reevesby Island tiger snake is about 5 times more toxic than the Indian cobra *Naja naja*.

# King Island and Tasmanian tiger snake

*Notechis ater humphreysi* (Worrell, 1963)
*(Highly venomous)*

**Description**  Blunt head is distinct from a robust body. Younger snakes may be slimmer and similar to other tiger snakes. Grows to 1.5 m in length. Dorsally, may be jet black, jet black with lighter crossbands, grey with black flecks forming faint bands or an un-banded grey or brown. The ventral surface is usually a lighter colour. Midbody scales are in 19, 17 or sometimes 15 rows, ventrals number 161–174, subcaudals 48–53 (single) and the anal scale is single.

**Habits and habitat**  On Christmas and New Year Islands, it feeds on mutton bird chicks. King Island snakes have been reported to be cannibalistic. In Tasmania, it feeds on mammals, birds, lizards and has been observed feeding on introduced trout. Occurs in a variety of habitats throughout its range, from coastal dunes, marshland, sclerophyll forests, coastal heath grasslands to agri-developed land. It is live bearing and produces large numbers of young. The venom from King Island tiger snakes is slightly more toxic than most mainland *N. scutatus*, but samples tested from Tasmanian and New Year Island tiger snakes is less toxic. There is one captive record (from Tasmania) of this species using constriction to restrain its prey whilst the venom immobilised it. The venom is neurotoxic and has a coagulation action.

# Chappell Island tiger snake

*Notechis ater serventyi* (Worrell, 1963)
*(Highly venomous)*

**Description**  Blunt head is distinct from a robust body. The giant of the tiger snakes, averaging 1.9 m in length. Dorsally, olive-brown to almost black, sometimes with lighter crossbands. The ventral surface is usually lighter in colour. Juveniles are banded. Midbody scales are in 17 rows, ventrals number 160–171, subcaudals 47–52 (single) and the anal scale is single.

**Habits and habitat**  Occurs on Chappell, Badger, Babel, Cat, Forsyth, Vansittart and Flinders Islands of the Furneaux Group. Usually diurnal. Adults feed almost entirely on migratory mutton birds, then fast for the rest of the year after the mutton birds have departed. The growth rate in captivity of this species has been found to be twice that of *N. scutatus*. Usually sluggish in disposition, and larger specimens can be approached without any apparent concern by the snake. It is live bearing, and up to 31 neonates have been recorded. The venom is the least toxic of all the tiger snakes. Its neurotoxicity is different to other tiger snakes and its toxicity is about 1.8 times more toxic than the Indian cobra *Naja naja*. The venom is neurotoxic but lacks the same lethal neurotoxin as other *Notechis* snakes. It also has a coagulant.

# Western tiger snake

*Notechis ater (scutatus) occidentalis* (Glauert, 1948)
*(Highly venomous)*

**Description** Head is distinct from a robust body. Grows to 2.0 m in length. Dorsally, steel-blue to black with bright yellow bands; unbanded specimens occur. The ventral surface is yellow, tending black towards the tail. Midbody scales are in 17 or 19 rows, ventrals number 140–165, subcaudals 36–51 (single) and the anal scale is single (rarely divided).

**Habits and habitat** Diurnal, crepuscular or nocturnal in hot weather. It is a live bearer and produces up to 90 young. Smaller clutches of about 14–20 are more normal. Feeds on frogs, lizards, birds, mammals and sometimes fish. A life span of 10 years has been measured. A mortality rate, with newly born snakes, of 90 per cent has been recorded on Carnac Island. Its occurrence is associated with moist temperate climates. Occurs near waterbodies such as swamps, creeks, rivers, lakes etc. Is essentially terrestrial, but has been known to climb bushes and small trees, both in seeking refuge from floods or in search of food. Occurs in large numbers in favourable times. Its venom is about 4 times more toxic than the Indian cobra *Naja naja* and is the second most toxic of all the tiger snake venoms. The venom is probably similar in action to *N. scutatus* but definitely has neurotoxins and coagulants.

# Common tiger snake

*Notechis scutatus* (Peters, 1861)
*(Highly venomous)*

**Description**  Flat blunt head, slightly distinct from a robust body. Body capable of being flattened along entire length when snake is agitated or basking. Average length 0.9 m, maximum length 1.2 m, but has been recorded at 2.0 m. Highly variable in colour, with base colours of brown, grey-olive, green with lighter crossbands usually of creamy yellow. Occasionally unbanded specimens are found. Scales appear like overlapping shields, especially around the neck. Ventrals number 140–190, subcaudals 35–65, midbodies in 17 or 19 rows and the anal scale is single.

**Habits and habitat**  Found either in wet temperate areas around swamps, mesic mountain slopes, along irrigation channels or adjacent large waterbodies. The reason for this is that frogs represent the most significant component of its diet (50 per cent). Other prey items include mammals (24 per cent), lizards (15 per cent), birds (9 per cent) and snakes (*N. scutatus* 2 per cent). Eels and fish have been recorded as occasional prey items. The average size of mainland tiger snakes is lower than the island *N. ater* species, reflecting a smaller average prey size on the mainland. Tiger snakes are particularly prone to parasitism — a range of roundworms, tapeworms and flukes have been found. The species' survival with such a parasite burden makes them truly remarkable. It is viviparous, producing litters from 12 to 80. Although it is still common in many areas, overall numbers have been reduced drastically, the main cause being habitat alteration. For example, river-level control along the Murray River has reduced annual flooding patterns, altering the watertable and reducing the number of swamps and lagoons. In south-western Victoria, the extensive stone walls, once a haven for snakes, have been replaced by wire fences. Lakes and swamps have been drained for farming purposes. In some developed areas, there are now reduced but balanced tiger snake populations. For instance, they are occasionally found in Sydney or Melbourne along watercourses. Tiger snakes are abundant in Altona, a Melbourne suburb. Fortunately, for their conservation, tiger snakes are well represented in national parks. Pooled samples of freeze-dried venoms from this species are less toxic than *N. a. ater*, *N. a. occidentalis* and *N. a. niger*. Its venom is 4.3 times more toxic than the Indian cobra *Naja naja*. The venom contains neurotoxins coagulants, haemolysins and myotoxins.

*Unbanded* N. scutatus.

# Taipan

*Oxyuranus scutellatus* (Peters, 1867)
*(Highly venomous)*

**Description** Large rectangular-shaped head, narrow neck, cylindrical body, small scales on the dorsal surface of neck and red eyes. Recorded up to 2.8 m in length; average length around 2.0 m. The scales are feebly keeled (especially on the neck and vertebral region) to smooth. Ventrals number 220–250, subcaudals 45–80, midbodies are in 21–23 rows and the anal scale is single. Fang length varies from 7.9–12.1 mm in adults. Always exhibits a pale or creamish colour around the head and lips, which is more predominant in juveniles. With age, the paleness recedes to just the lip area. Dorsally, colour varies from light brown, dark brown to black, coppery-red to olive.

**Habits and habitat** A master at hunting. Strikes with amazing speed, more than once in some cases, and then holds back and waits (snap-bite and release) until the prey quickly succumbs to the venom. This enables the snake to avoid injuries from its prey. Alert and nervous when approached by humans, and extremely dangerous if threatened. In most instances, it will move away undetected, but if cornered it will readily defend itself. The venom is 7.8 times more toxic (in mice) than the Indian cobra *Naja naja*. In one instance, a young boy died less than an hour after receiving 12 bites from a taipan. The venom contains a number of neurotoxins, one of which is also muscle-destroying, a procoagulant and a number of other toxins. Feeds predominantly on rats, especially *Melomys* (spp.). Prefers warm-blooded animals as prey — birds, mice and bandicoots are recorded as food items. Lizards are also taken. Sexually mature when it attains a length of about 1.5 m, which in captivity takes about 16 months for males and 28 months for females. Oviparous, producing 7–20 eggs. Occurs in the wetter coastal areas of Qld, NT and northern WA. Inhabits the sugar cane areas of Queensland, where rats abound, and this industry is thought to have resulted in increases in its populations. The effects of new pesticides (used for rodent control in the sugar industry) on taipans is unknown.

# Inland taipan, western taipan, small-scaled snake or fierce snake

*Oxyuranus (Parademansia) microlepidotus* (McCoy, 1879)
*(Highly venomous)*

*Summer colour.*

**Description** Head is long and oblong-shaped, but to a lesser extent than *O. scutellatus* and with a greater slope from the frontal to the rostral scales. The scales on the neck are small. The head is slightly distinct from a robust body. The neck is narrow, but more robust than *O. scutellatus*. The eye colour is black and smaller in proportion to *O. scutellatus*. The fang length varies from 3.5–6 mm in adults. Grows to 2.7 m in length. Dorsally, varies from dark brown to a light straw colour, with a series of darker flecks arranged in faint crossbands. There are dramatic changes from winter to summer colouration, being darker in winter and lighter in summer. The head is sometimes black or darker than the rest of the body. The ventral surface is a rich cream to yellow, sometimes with orange blotches. Midbody scales are in 23 rows, ventrals number 211–250, subcaudals 52–70 (mostly or all divided; a small number close to the anal scale may be single) and the anal scale is single.

**Habits and habitat** Occurs in inland regions near the junction of Qld, NSW, NT and SA. The original specimens were collected from the junction of the Darling and the Murray Rivers before 1879, but since then none have been collected anywhere near that site. Another specimen was collected about that time at Bourke in NSW, but, again, no other specimens have been collected in that area. During the Christmas period of 1989, a snakebite occurred near Mildura (not far from the junction of the Darling and the Murray Rivers) and subsequent testing of the patient with a Commonwealth Serum Laboratories venom detection kit proved positive for taipan antivenom. This indicates that the species could still be in this area. Its environment is one of harsh extremes of temperature and rainfall. Normally diurnal, its movements are restricted to times during the day when temperatures permit activity, usually early morning or before sunset. Confined to the Lake Eyre drainage system and peripheral areas, such as watercourses, flood plains and adjacent gibber plains. Seeks refuge in solution holes and cracks in the ground caused by the drying of flood plains. Feeds almost exclusively on the long-haired rat *Rattus villosissimus*, but has been recorded taking *Mus musculous*, *Antichinomys* (sp.) and a *Dasyurid* (sp.). Like *O. scutellatus*, when hunting it uses a snap-bite and release method, and holds off while the prey quickly succumbs to the potent venom. It is an egg layer, producing 12–20 eggs. Mating occurs in January, February, March, September, October and December. The most toxic terrestrial snake venom known, being 50 times more toxic than the Indian cobra *Naja naja*. It contains neurotoxic, procoagulant and myotoxic activities.

*Winter colour.*

# King brown or mulga snake

## *Pseudechis australis* (Gray, 1842)
### *(Highly venomous)*

**Description**  Head is slightly distinct from a robust body. Grows to 2.7 m in length. Dorsally, may be light brown to tan, coppery-red, dark brown to black scales (with the leading half of the scale yellow, dark brown scales with the first 10 per cent of the leading edge yellow). Northern specimens are sandy brown; central deserts specimens are strongly 'speckled', with a white spot on each scale; and southern populations tend to be darker. Scientific evidence on chromosomes, scalation, general morphology and electrophoretic patterns of blood proteins supports the retention of these populations in *P. australis*. The ventral surface is cream to white, sometimes with orange blotches. Midbody scales are in 17 rows, ventrals number 185–225, subcaudals 50–75 (all single or all divided or anteriorly single and posteriorly divided) and the anal scale is divided.

**Habits and habitat**  Either crepuscular or nocturnal; rarely diurnal. Distributed widely throughout Australia. Occurs in all states except Victoria and Tasmania, and in most habitat types except alpine and cool wet environments. Although an early record reported this species as live bearing, all recent captive breedings have reported oviparity, with records of 11–16 eggs in a clutch. The authors record that populations from Eyre Peninsula in South Australia show a skewed sex ratio in favour of males. Feeds on lizards, other species of snake, mammals, birds and frogs. Its venom differs from most other Australian dangerous elapids in that whilst it has been shown to possess slight neurotoxicity, it is not significant. The main toxin is a myotoxin or muscle-destroying poison. Its venom is also strongly anticoagulant in action. Overall toxicity is only about

25 per cent as toxic as that of the Indian cobra *Naja naja*. The sex chromosomes, scalation, blood proteins, blood coagulation inhibitors, oviparity, robust body, arid habitat and other morphological features of this species most closely associate it with *P. butleri*. Male combat has been recorded with this species.

# Butler's snake

*Pseudechis butleri* (Smith, 1982)
*(Highly venomous)*

**Description** Previously considered to be a variant of *Pseudechis australis*. Head is slightly distinct from a robust body. Grows to 1.6 m in length. Dorsally, black or dark grey, with many scales exhibiting a prominent cream to bright yellow blotch. The top of the head and nape are dark, lacking scales with blotches, and it has a reddish-brown tinge on the snout and sides. The ventral surface is cream to yellow, with black flecks and an uneven black edge at the base. Juveniles differ from adults, having a dark bluish-grey dorsal colour and being obscurely patterned, with pale green markings on the ventral surface. Midbody scales are in 17 rows, ventrals number 204–216, subcaudals 55–65 (anteriorly single and posteriorly divided) and the anal scale is divided.

**Habits and habitat** Either nocturnal or diurnal. Occurs in stony soils associated with mulga woodlands and shrublands. Shelters in abandoned burrows, weathered caves and beneath dead vegetation. It is an egg layer and produces 9–12 eggs. Feeds on small mammals and reptiles. Nothing certain is known about the venom, which is assumed to be similar to *P. australis* venom until further studies are carried out. The sex chromosomes, scalation, blood proteins, blood coagulation inhibitors, oviparity, robust body, arid habitat and other morphological features of this species most closely associate it with *P. australis*.

# Collett's snake

*Pseudechis colletti* (Boulenger, 1902)
*(Highly venomous)*

**Description** Head is slightly distinct from a robust body. Grows to 2.0 m in length. Dorsally, dark brown to black, with lighter brown to red scales forming broken bands, which merge into the brighter colours on the flanks. The head tends to be darker. The ventral surface is cream, pink to orange. Juveniles are more brightly coloured than adults. Midbody scales are in 19 rows, ventrals number 215–235, subcaudals 50–70 (anteriorly single and posteriorly divided) and the anal scale is divided.

**Habits and habitat** Diurnal, often active after rains. Occurs in open black soil plains of central Queensland, where it seeks refuge in dried soil cracks. It is an egg layer and produces 6–19 eggs. Probably feeds on lizards and mammals. When annoyed, it inflates its body and hisses loudly. The venom has been shown to possess myolytic activity and is about 24 per cent as toxic as the Indian cobra *Naja naja*. Combat has been observed between males in captivity at Venom Supplies Laboratory. Strangely, it most closely resembles *P. papuensis* from New Guinea in respect to morphological and scale features.

# Blue-bellied or spotted black snake

*Pseudechis guttatus* (De Vis, 1905)
*(Highly venomous)*

**Description** Head is slightly distinct from a robust body. Grows to 2.0 m in length. Dorsally, may be glossy black, sometimes with scattered cream-tipped scales; occasionally mostly cream-coloured scales with scattered black scales. The ventral surface is grey to bluish-grey, sometimes with cream blotches. Midbody scales are in 19 rows, ventrals number 175–205, subcaudals 45–65 (anteriorly single and posteriorly divided) and the anal scale is divided.

**Habits and habitat** Predominantly diurnal, becoming nocturnal in hot weather. Occurs on rocky hill slopes, river flood plains, coastal forests, dry sclerophyll forests and woodlands. Seeks refuge in fallen timber, abandoned burrows and ground cracks. It is an egg layer and produces 7–13 eggs. Feeds on mammals, reptiles and frogs. When annoyed, it arches the forebody, flattens its neck and hisses loudly. The venom is haemolytic, coagulant and neurotoxic, and is about 26 per cent as toxic as the Indian cobra *Naja naja*. Male combat has been recorded for this species.

# Red-bellied black snake

*Pseudechis porphyriacus* (Shaw, 1794)
*(Highly venomous)*

**Description**  Head is slightly distinct from a robust body. Grows to 2.1 m in length. Dorsally, purplish-black with brilliant red, crimson, orange and occasionally white sides adjacent to ventral scales. The ventral surface is red, crimson, orange, grey and occasionally white. The subcaudals are black. Midbody scales are in 17 rows, ventrals number 170–215, subcaudals 40–65 (first third single and remainder divided) and the anal scale is divided.

**Habits and habitat**  Mostly diurnal. Usually associated with streams, rivers, creeks, swamps, lakes or generally a wet area. It is the only live bearer from the *Pseudechis* genus and produces 8–40 young. Being ovoviparous, possessing different electrophoretic properties in blood proteins, having cytogenetic differences, different body proportions, scalation and ecological differences, suggest it is divergent from other species of this genus. Feeds mainly on frogs, but will take lizards, mammals, birds and even fish. Normally shy and quick to retreat if approached. May flatten out its neck and body, and hiss loudly. Its venom is neurotoxic, mildly coagulant and myotoxic, and is about 22 per cent as toxic as the Indian cobra *Naja naja*. Male combat has been observed in this species.

# Dugite

*Pseudonaja affinis affinis* (Günther, 1872)
*(Highly venomous)*

**Description**  Small head, indistinct from a long slender body. Grows to 2.0 m in length. Dorsally, dark brown, olive or grey, usually with speckled irregular black scales. The ventral surface is whitish or creamy yellow, sometimes with darker flecks and orange blotches. Juveniles often have black markings around the head and nape, and a series of black markings on the body, forming a herringbone pattern. Midbody scales are in 19 rows, ventrals number 190–230, subcaudals 50–70 (divided) and the anal scale is divided.

**Habits and habitat**  Large, fast-moving and alert. Mainly diurnal, sometimes nocturnal during hot weather. Found over a wide variety of habitats throughout its range, including coastal dunes, semi-arid shrublands and woodlands, and wet sclerophyll forests. Seeks refuge in logs, abandoned burrows, fallen debris and under rocks. It is an egg layer and produces 13–20 eggs. Feeds on mammals, birds, lizards and frogs. As with most of the *Pseudonaja* genus, it catches and kills its prey by a combination of venom and constriction. When provoked, it raises its body in an S-shape, flattens its neck like a cobra and hisses loudly, while striking repeatedly. Male combat has been observed in this species. The venom is strongly neurotoxic, coagulant and myolytic. One of its neurotoxic components can be used to diagnose the neuromuscular disease myasthenia gravis. Occasionally, the venom may cause thrombocytopenia. It is about 90 per cent as toxic as the Indian cobra *Naja naja*.

**Subspecies**

*Pseudonaja affinis tanneri* (Worrell, 1961)
Similar in most respects to *P. affinis affinis*, but smaller and uniformly darker in colour. Grows to 2.0 m in length. Occurs on some of the islands of the Recherche Archipelago off the southern coast of Western Australia. Another population on Rottnest Island, near Perth, may be the same subspecies.

# Speckled brown snake

*Pseudonaja guttata* (Parker, 1926)
*(Highly venomous)*

**Description**  Small head, indistinct from a slender body. Grows to 1.4 m in length. Dorsally, mainly a straw-yellow to apricot, with black markings on many of the lower concealed edges of the scales. The black markings become more visible when the snake moves or inflates its body. Some specimens have broad darker bands, which gradually merge into the ground colour on either side. The throat and most of the labial scales are white. The ventral surface is white, creamish or yellow, with many orange blotches. Midbody scales are in 19 or 21 rows, ventrals number 190–220, subcaudals 44–70 (divided) and the anal scale is divided.

**Habits and habitat**  Diurnal, restricted mainly to black soil plains, sheltering in deep earth cracks, especially near water. Many specimens have parts of their tails missing, due possibly to predation when they leave part of their tail exposed after entering earth cracks. It is an egg layer. In captivity, it shows a preference for lizards over mammals. When harassed, it raises its forebody and flattens its neck like a cobra. Its venom is more toxic than that of the dugite, nearly as toxic as the western brown snake and about 1.6 times more toxic than the Indian cobra *Naja naja*. An unusual and unexpected predator is the smaller snake *Suta suta*, which has been recorded preying on *Pseudonaja guttata* at Goyder's Lagoon, in the far north-east of South Australia. Uses constriction as a method of prey restraint as well as venom to immobilise its prey.

# Peninsula brown snake

*Pseudonaja inframacula* (Waite, 1925)
*(Highly venomous)*

**Description** Head is glossy and indistinct from a slender body. Grows to 1.5 m in length. Dorsally, colour varies from dark brown to almost black, light yellow to chocolate, with or without some or numerous black scales. Banded specimens occur. The ventral surface is light to dark grey, with northern specimens tending to a mottled grey. Midbody scales are in 17 rows, ventrals number 190–227 (southern specimens tend to have lower ventral counts), subcaudals 52–62 (divided) and the anal scale is divided.

**Habits and habitat** Mostly diurnal, essentially being a sun-loving snake. Its darker colour allows it to bask on cooler days than most other members of the genus. Occurs in dry sclerophyll woodlands, limestone outcrops, coastal sand dunes and heaths. Occurs on Wardang Island, where it grows to a large robust size. It is an egg layer and produces 12–20 eggs. Feeds on lizards (even *Tiliqua rugosa*), mammals, birds and frogs. Less inclined to bite and tends to be less nervous than other members of this genus. Nothing is known about the venom, but numerous bites have been attributed to this species, all of which have responded well to brown snake antivenom treatment. Specimens found in developed areas are larger than those found in natural habitats. This suggests that they have adapted to feeding on abundant supplies of introduced mice *Mus musculous*. Uses constriction to restrain prey whilst the venom immobilises it.

99

# Ingram's brown snake

*Pseudonaja ingrami* (Boulenger, 1908)
*(Highly venomous)*

**Description**  Head is indistinct from a slender body. Grows to
1.76 m in length. Dorsally, it has 5 colour forms: glossy black-brown;
dark brown anteriorly, golden-brown posteriorly; uniform golden-
brown; head and nape grey-brown to dark brown, body light to rich
yellow-brown; and pale olive-brown. The scales on all forms are
dark on the tips. On the lighter forms, the head is darker. The ventral surface is pale to bright yellow to orange, often fading to cream
on the chin. A series of 2 orange dots are present on each side of the
ventral scales, increasing to 4 or more posteriorly. The eyes are dark
orange-brown, appearing to be black. Its buccal cavity is predominantly black. Midbody scales are in 17 rows, ventrals number
190–223, subcaudals 55–72 (divided) and the anal scale is divided.
Differs from *P. textilis* in having 7 infralabial scales instead of 6, and
darker eyes and darker buccal cavity.

**Habits and habitat**  Diurnal and seeks refuge in earth cracks on
black soil plains which are subject to seasonal flooding. Feeds
predominantly on mammals, especially the long-haired rat *Rattus
villosissimus* and introduced mouse *Mus musculous*. Active mostly during early morning. Many individuals can be found with parts of their
tails missing. It is believed this is due to predation when the snakes
leave their tails partially exposed after entering earth cracks. When
harassed, it raises its forebody and flattens its neck like a cobra.
Nothing is known of its venom toxicity, although it is expected to be
similar to *P. textilis*. Uses constriction to restrain prey whilst the
venom immobilises it.

# Ringed brown snake

*Pseudonaja modesta* (Günther, 1872)
*(Venomous)*

**Description**  Head is indistinct from a slender body. Grows to 0.6 m in length. Dorsally, olive-tan or a rich reddish-brown, with a series of evenly spaced narrow black crossbands numbering 4–12, which can fade or disappear with age. The head and nape may be black and divided by a cream band or the head may be similar in colour to the body, with black markings around the eye, snout and nape. The ventral surface is creamish with orange or grey blotches. The iris is orange-brown. Midbody scales are in 17 rows, ventrals number 145–175, subcaudals 35–55 (divided) and the anal scale is divided.

**Habits and habitat**  The smallest of the *Pseudonaja* genus. Diurnal, but may become nocturnal during hot weather. Occurs in the semi-arid to arid parts of its range, where it can be found along dry watercourses, and in old animal burrows, particularly those of the central netted dragon *C. tenophorus nuchalis*. It is an egg layer and produces 7–11 eggs. Feeds on small diurnal lizards. When provoked, it raises its body in an S-shape and hisses. Uses constriction to restrain prey whilst the venom immobilises it.

# Western brown snake or gwardar

*Pseudonaja nuchalis* (Günther, 1858)
*(Highly venomous)*

**Description** Head is indistinct from a slender body. Grows to 1.5 m in length. Dorsally, the colour is highly variable and it would be impossible to list all the variants; however, the ground colour varies from tan, light brown, dark brown to black, coppery-red and straw-yellow. Superimposed on the ground colour, any combination of black spots, blotches and banding is possible. Young snakes hatched in captivity from the one clutch exhibit many colour variations. In addition, colour intensity darkens in the winter. Many specimens have a dark nuchal band. The ventral surface may be a mottled grey or yellow, sometimes with orange spots. Midbody scales are in 17 (rarely 19) rows, ventrals number 180–230, subcaudals 50–70 (divided) and the anal scale is divided. The rostral scale extends further back over the head in this species than with other *Pseudonaja*, and when viewed from above it appears chisel-like. The iris is red. The head is more glossy in this species than other *Pseudonaja*.

**Habits and habitat** Mainly diurnal, but becomes nocturnal on some hot nights. Widespread throughout most mainland states and inhabits most habitat types. It is an egg layer, producing 13–22 eggs. Feeds on mammals, birds, lizards and frogs. Its threat pose involves an S-shaped posture, with the neck flattened like a cobra. Male combat has been observed in this species. The venom is powerfully neurotoxic, procoagulant in activity, and is 1.5 times more toxic than the Indian cobra *Naja naja*. Uses constriction to restrain prey whilst the venom immobilises it.

# Common brown or eastern brown snake

*Pseudonaja textilis* (Duméril, Bibron and Duméril, 1854)
*(Highly venomous)*

**Description** Head is indistinct from a slender body. Grows to 2.5 m in length. Dorsally, may be tan, grey, light brown, dark brown to nearly black and straw-yellow. Occasionally, banded specimens occur, but usually banding is restricted to some juveniles, where it can be quite marked. The head is not as glossy as *P. nuchalis* or *P. inframacula*. The ventral surface is cream, yellow or brown, heavily blotched with orange or dark grey. Midbody scales are in 17 rows, ventrals number 185–235, subcaudals 45–75 (divided) and the anal scale is divided.

**Habits and habitat** Normally diurnal. Fast-moving, extremely alert, nervous and ill-natured if cornered. Probably the only Australian snake that will sometimes attack if disturbed. Herpetologists find it the most difficult to handle. Northern specimens tend to be larger and will defend themselves more aggressively than southern specimens. Found over a wide range of habitats, from temperate sclerophyll forests to arid deserts. Probably has been the most successful Australian elapid in adapting to modified environments. It is an egg layer and produces 10–30 eggs. Feeds on lizards, mammals, birds and frogs. Its threat pose involves raising its forebody in an S-shape and hissing loudly. Male combat has been observed in this species. The venom contains at least 2 neurotoxins, is strongly coagulant, weakly haemolytic and is 12 times more deadly than the venom of the Indian cobra *Naja naja*. Uses constriction to restrain prey whilst the venom immobilises it.

# Müller's snake

*Rhinoplocephalus bicolor* (Müller, 1885)
*(Venomous)*

**Description** Head is slightly distinct from a moderately robust body. Grows to 0.4 m in length. Dorsally, olive-grey to light olive-grey, merging to a lighter yellowish-orange on the sides. The head is slightly darker than the body. The ventral surface is white or creamish. Midbody scales are in 15 rows, ventrals number 145–165, subcaudals 25–35 (single) and the anal scale is single.

**Habits and habitat** Crepuscular in nature, it is found in sandy areas subject to inundation. Seeks refuge in disused ant nests (*Iridomyrex conifer*), under granite slabs and fallen grass trees *Xanthorrhoea* spp. in the vicinity of swamps. Diet consists mainly of skinks, but frogs have also been recorded as prey items. It is an egg layer and 1–4 hatchlings have been recorded in late summer to May. When threatened, it raises its head and hisses. Active throughout the year. At maturity, males are the same size as females.

# Carpentaria whip snake

*Rhinoplocephalus boschmai (Unechis carpentariae)*
(Brongersma & Knaap-van Meeuwen, 1961)
*(Venomous)*

**Description** Head is slightly distinct from a moderately robust body. Grows to 0.45 m in length. Dorsally, yellow-orange to brown, sometimes with a darker vertebral colour. The side of the head and forebody is a lighter shade of dorsal colour. The ventral surface is white. The midbody scales are in 15 rows.

**Habits and habitat** Found on hard soils with open vegetation types. Feeds on small lizards. Active throughout the year.

# Short-tailed snake

*Rhinoplocephalus (Unechis) nigriceps* (Günther, 1863)
*(Venomous)*

**Description** Head is slightly distinct from a moderately robust body. Grows to 0.6 m in length. Dorsally, dark brown to black, with a darker vertebral stripe or zone. Laterally, reddish to purplish-brown, merging into the vertebral zone. The head is black on top and is continuous with the vertebral stripe. The lips and ventral surface are white. Midbody scales are in 15 rows, ventrals number 152–164, subcaudals 23–29 and the anal scale is single.

**Habits and habitat** Nocturnal. Found mainly in mallee-type habitats, often on or near hill slopes and rocky outcrops. It is a live bearer and has been recorded with a litter of 3–4 young. Feeds on small lizards, other small elapids and blind snakes. Active throughout the year.

## *Rhinoplocephalus spectabilis* (Krefft, 1869)
### *(Venomous)*

**Description**  Head is slightly distinct from a moderately robust body. Grows to 0.4 m in length. Dorsally, brick red or light grey-brown, with a darker spot at the base of each scale. The top of the head and nape are glossy black. The tip of the snout has a dark brown or black stripe. Between the snout and the dark head a lighter colour similar to the dorsal colour extends across the head and through the eyes and lips. The ventral surface is white. Midbody scales are in 15 rows, ventrals number 135–168, subcaudals 21–36 (single) and the anal scale is single.

**Habits and habitat**  Nocturnal and sometimes diurnal in winter. Found in open woodlands, myall saltbush associations and eucalypts with heavy and sandy soils. Seeks refuge in abandoned holes and surface rubbish. Often found more in winter warming in the sun under thin layered refuge. It is a live bearer. Feeds on small lizards. Active throughout the year.

**Subspecies**

***Rhinoplocephalus spectabilis nullarbor*** (Storr, 1981)
Differs from *R. s. spectabilis* in having reduced markings on the head and a longer tail. Inhabits limestone-based soils supporting chenopod shrublands.

# Northern desert banded snake

*Simoselaps anomalus* (Sternfeld, 1919)
*(Venomous)*

**Description**  Head is indistinct from a robust body. The eyes are small. The snout is rounded. Grows to 0.21 m in length. Dorsally, yellow or orange with 24–40 black bands, which are narrower than the lighter interspaces, extending from the nape to the tail. The band across the nape is wider than the other bands. The head is black with a narrow band across the snout, level with the nostrils. The ventral surface is a paler shade of the dorsal colour. The snout is shovel-shaped and overhangs the bottom lip. The tail is short. Mid-body scales are in 15 rows, ventrals number 119–130, subcaudals 17–27 (divided) and the anal scale is single.

**Habits and habitat**  Burrowing and nocturnal. Found on and between sand dunes vegetated with hummock grasses. Feeds on burrowing skinks (*Lerista spp.*), which it constricts in a series of tight coils. It is egg laying. Adult females are larger than males.

## *Simoselaps approximans* (Glauert, 1954)
### *(Venomous)*

**Description** Head is indistinct from a robust body. Grows to 0.37 m in length. Dorsally, grey to grey-brown, with 51–96 narrow (less than one scale wide) white bands from the nape to the tail tip. The head is greyish-brown on top with a band across the snout level with the nostrils, continuous on the sides. The eye is small. The ventral surface is cream. The upturned snout is shovel-shaped and overhangs the bottom lip and has an acute cutting edge. The tail is short. Midbody scales are in 17 rows, ventrals number 158–181, subcaudals 19–27 (single) and the anal scale is divided.

**Habits and habitat** Favours heavy stony soils associated with *Acacia* woodlands. Burrows into soil cracks, insect holes, beneath stumps and rocks. Feeds on lizard eggs. It is egg laying. Adult males are the same size as the females.

# Australian coral snake

*Simoselaps australis* (Krefft, 1864)
*(Venomous)*

**Description**  Head is indistinct from a small robust body. Snout is shovel-shaped, with a sharp cutting edge. Grows to 0.5 m in length. The tail is short. Dorsally, pink to brick red, with 50–60 narrow ragged-edged crossbands formed of pale cream-centred scales, with dark edges extending from behind the nape to the tip of the tail. A broad pale-edged black blotch is situated on the nape of and another on the head, extending from eye to eye. The ventral surface is cream. Midbody scales are in 17 rows, ventrals number 140–170, subcaudals 15–30 (divided) and the anal scale is divided.

**Habits and habitat**  Burrowing and nocturnal. Seeks refuge during the day under rocks and logs in woodlands, shrublands and hummock grasslands. Red sandy soils are favoured. It is egg laying and produces 4–6 eggs. Feeds on small lizards and their eggs. Adult females are larger than the males.

# Desert banded or Jan's banded snake

*Simoselaps bertholdi* (Jan, 1859)
*(Venomous)*

**Description**  Head is indistinct from a robust body. The snout is depressed and rounded. The tail is short. Grows to 0.33 m in length. Dorsally, yellow to reddish-orange, with 18–31 black bands from the nape to the tip of the tail. Darker edges occur on the lighter scales, but disappear laterally. The eye is small. The head is white to pale grey, densely flecked with blackish-brown. The ventral surface is pale pink, white to yellow, sometimes with mid-ventral blotches between the black rings. Midbody scales are in 15 rows, ventrals number 112–131, subcaudals 15–25 (divided) and the anal scale is divided.

**Habits and habitat**  Occurs in mallee and myall woodlands, heathlands, coastal dunes and rocky outcrops. Nocturnal and sometimes diurnal, even in hot weather. Seeks refuge mainly in thick leaf litter or loose soil beneath trees or shrubs, where it feeds on small burrowing lizards such as *Lerista* spp. and their eggs. It is an egg layer and produces 1–8 eggs. Adult females are larger than the males.

111

# Narrow banded burrowing snake

*Simoselaps fasciolatus fasciolatus* (Günther, 1872)
*(Venomous)*

**Description** Head is indistinct from a robust body. The snout is upturned and shovel-shaped. The tail is short. Grows to 0.4 m in length. Dorsally, white, cream to pale pink, with numerous narrow ragged-edged dark brown to black crossbands from the nape to the tail tip. There is a dark blotch on the head and another on the nape. The ventral surface is white. Midbody scales are in 17 rows, ventrals number 140–175, subcaudals 15–30 (divided) and the anal scale is divided.

**Habits and habitat** Nocturnal and prefers sandy soils, where it seeks refuge in soil cracks, insect holes and under rocks. It is an egg layer. Feeds on small burrowing lizards and their eggs. Adult females are larger than males.

**Subspecies**

***Simoselaps fasciolatus fasciata*** (Stirling & Zietz, 1893)
Differs from *Simoselaps fasciolatus fasciolatus* in having a narrower nuchal band.

## *Simoselaps incinctus* (Storr, 1968)
### *(Venomous)*

**Description**  Head is indistinct from a robust body. The snout is upturned and shovel-shaped, with a sharp cutting edge. It has a short tail. Grows to 0.3 m in length. Dorsally, tan to reddish-brown, fading on the sides, with each scale dark edged. There are dark black blotches on the head and nape. The ventral surface is cream. Mid-body scales are in 17 rows, ventrals number 140–165, subcaudals 18–30 (divided) and the anal scale is divided.

**Habits and habitat**  Found in clay, loamy and stony soils with vegetation of woodlands, shrublands and hummock grasslands. Shelters under rocks, logs and other debris or in unused insect holes. A nocturnal and burrowing type. It is an egg layer. Feeds on small burrowing lizards.

# Coastal burrowing snake

*Simoselaps littoralis* (Storr, 1968)
*(Venomous)*

**Description** Head is indistinct from a robust body. Grows to 0.4 m in length. Snout is depressed and rounded. It has a small tail and small eye. Dorsally, pale yellow to orange, with 20–42 narrow black crossbands from the nape to the tip of the tail. The head is white, flecked with black, with a distinct black bar across the back of the head. The ventral surface is whitish. Midbody scales are in 15 rows, ventrals number 104–125, subcaudals 16–23 (divided) and the anal scale is divided.

**Habits and habitat** Occurs on coastal dunes and limestones vegetated with beach spinifex and shrubs, occasionally extending into adjacent hummock grasslands. Nocturnal and burrowing in habit. It is an egg layer. Feeds on small burrowing lizards. Adult females are larger than males.

## *Simoselaps minimus* (Worrell, 1960)
### *(Venomous)*

**Description**  Known only from two specimens. Head is indistinct from a robust body. Grows to 0.22 m in length. The snout is depressed and rounded. It has a short tail. Its eyes are small. Dorsally, creamish with each scale edged dark brown. Has a black bar on the head and neck. The ventral surface is white. Midbody scales are in 15 rows, ventrals number 125–127, subcaudals 19–22 (mostly divided) and the anal scale is divided.

**Habits and habitat**  One specimen found in a coastal environment. Little is known about this snake, and it is assumed to be similar to other members of the *Simoselaps* genus. Adult females are larger than males.

# Half-girdled snake

*Simoselaps semifasciatus semifasciatus* (Günther, 1863)
*(Venomous)*

**Description** Head is indistinct from a robust body. Grows to
0.35 m in length. Snout protrusive, upturned, and shovel-shaped,
with a sharp cutting edge. Dorsally, light brown to reddish, with
45–80 dark crossbands. There are broad dark bands across the head
and the nape. The ventral surface is cream. Midbody scales are in 17
rows, ventrals number 145–190, subcaudals 14–26 (divided) and
the anal scale is divided.

**Habits and habitat** Nocturnal. Occurs in mallee woodlands,
arid woodlands and other mesic habitats. Seeks refuge under logs
and stones. It is an egg layer and feeds only on reptile eggs. Adult
females are larger than males. Uses constriction to restrain prey
whilst the venom immobilises it.

*sub species pg. 118*

# Half-girdled snake

## Subspecies

### *Simoselaps semifasciatus roperi* (Kinghorn, 1931)

Differs from *S. semifasciata semifasciata* in having less dark bands (38–74) and the width of the bands is greater. Adult females are larger than males.

# Robust burrowing snake

*Simoselaps warro* (De Vis, 1884)
*(Venomous)*

**Description**  Its inclusion in the *Simoselaps* genus is only tentative. Head is indistinct from a robust body. Has a depressed snout, lacking cutting edge. Grows to 0.4 m in length. Dorsally, orange to orange-brown, with darker scale edges presenting a mesh-like appearance. Has a cream head, densely flecked with black, and a black nuchal band. The ventral surface is creamish white. Midbody scales are in 15 rows, ventrals number 135–165, subcaudals 15–25 (divided) and the anal scale is divided.

**Habits and habitat**  A nocturnal burrowing snake. Seeks refuge in leaf litter and under logs in dry sclerophyll forests and woodlands, associated with grasslands. It is an egg layer. Feeds on small lizards. Adult females are larger than males.

# Ord curl snake

*Suta (Denisonia) ordensis* (Storr, 1984)
*(Venomous)*

**Description** Broad depressed head distinct from a robust body. Grows to 0.75 m in length. Dorsally, brown to greyish-brown, with each scale finely dark edged. The head is darker on top and even black in juveniles. The ventral surface is whitish, flushed with grey on each anterior edge of ventral and subcaudal scales. Midbody scales are in 19 rows, subcaudals are all single and the anal scale is single. Differs from *Suta suta* in having duller, less distinguishable or absent facial patterns.

**Habits and habitat** Nocturnal. Seeks refuge in ground cracks of flood plains. It is a live bearer. Probably feeds on small lizards.

# Curl or myall snake

*Suta (Denisonia) suta* (Peters, 1863)
*(Venomous)*

**Description** Broad depressed head is distinct from a robust body. Grows to 0.9 m in length. Dorsally, may be light brown to tan, dark brown, brick red or olive-green. The head is usually darker, with a dark line extending from the snout through the eyes to the nape. The supralabials to the eye are often lighter. Lips and ventral surface are cream with grey markings on the chin. Midbody scales are in 19 or 21 rows, ventrals number 150–170, subcaudals 20–35 (single) and the anal scale is single.

**Habits and habitat** Mostly nocturnal and sometimes diurnal. Seeks refuge under fallen timber, stones and in earth cracks in a wide range of habitats, from arid deserts, flood plains, mallee woodlands, myall saltbush associations to rocky outcrops. It is a live bearer and litters of 6 young have been recorded. Feeds on small lizards, mammals and frogs, and has been recorded eating another juvenile elapid, *Pseudonaja guttata*, in the Goyder's Lagoon area of north-eastern South Australia. Has a spectacular defence posture; it flattens its whole body and presents a tight loop or curl and will strike or feign a strike at any potential threatening movement.

# Rough-scaled or Clarence River snake

*Tropidechis carinatus* (Krefft, 1863)
*(Highly venomous)*

**Description** Head is distinct from a robust body. Grows to 1.0 m in length. All dorsal scales are strongly keeled. Dorsally, olive-green to brown, with a series of black bands, which can be distinct or broken over the body and absent on the tail. The black bands are more prominent in juveniles. Paler to white around the lips. The ventral surface is creamy-yellow to olive-green, often with darker blotches. Midbody scales are in 23 rows, ventrals number 160–185, subcaudals 50–60 (single) and the anal scale is single.

**Habits and habitat** Mainly nocturnal and sometimes diurnal; partly arboreal. Seeks refuge in dense fallen timber, earth cracks and other cavities in rainforests or areas adjacent to watercourses and swamps. It is a live bearer and produces 5–18 young in late summer and autumn. Females only reproduce in alternate years or greater. Adult males and females are the same size. Main diet consists of mammals (48 per cent) and frogs (41 per cent). It also has been recorded taking small lizards and birds. Normally inoffensive and very rarely seen, but will defend itself vigorously if threatened. Often ascends into low bushes or trees (sometimes banana trees) to feed and bask. Herpetologists are very wary of it in captivity of its nervous but pugnacious nature when threatened. A medical summary of 12 snakebites attributed to this species claims that 9 of those were by snakes kept in captivity. There is only one definite fatality recorded, but a number of other fatalities could possibly be due to this species. Its venom is powerfully neurotoxic, strongly coagulant and myotoxic or muscle destroying, and is about half as toxic as the Indian cobra *Naja naja*. The venom is closely related to *N. scutatus* venom. It is often confused with the harmless snake the keelback *styporynchus mairii*.

# Dwyer's snake

*Unechis (Rhinoplocephalus) dwyeri* (Worrell, 1956)
*(Venomous)*

**Description** Head is slightly distinct from a moderately robust body. Grows to 0.5 m in length. Dorsally, reddish to brownish, and on some each scale has a fine dark edge, giving a netted appearance. The top of the head is black, with paler areas around the eyes. The labials and ventrals are whitish to yellowish. Midbody scales are in 15 rows, ventrals number 147–152, subcaudals 25–34 (single) and the anal scale is single.

**Habits and habitat** Nocturnal and favours well-timbered areas and rocky outcrops, where it shelters under fallen timber, rocks, dead vegetation and in soil cracks. It is a live bearer, and mating has been observed in spring and autumn. Feeds on small lizards and smaller snakes. In favourable areas, it congregates in small colonies.

# Little whip snake

*Unechis (Rhinoplocephalus) flagellum* (McCoy, 1878)
*(Venomous)*

**Description** Head is barely distinct from a moderately slender body. Grows to 0.4 m in length. Dorsally, dull greyish-brown, fading on the flanks adjacent to the ventrals. Each scale is darker at the base. The top of the head is black, extending to the nape, with pale areas around the eyes and cheeks. A pale bar across the snout divides the head. The ventral surface is creamish. Midbody scales are in 17 (rarely 15) rows, ventrals number 125–150, subcaudals 20–40 (single) and the anal scale is single.

**Habits and habitat** Nocturnal. Sometimes confused with juvenile brown snakes *Pseudonaja textilis*. Found in highland and lowland woodlands, dry sclerophyll forests, where it seeks refuge under rocks, logs and loose sandy soils. It is a live bearer and produces up to 4 young in September, February and March. Adults have been found in pairs. Feeds on small lizards and insects. If disturbed, it flattens its body into tight coils. Not considered dangerous, but its bite can cause swelling of the lymph nodes.

# Gould's or black-headed snake

*Unechis (Rhinoplocephalus) gouldii* (Gray, 1841)
*(Venomous)*

**Description**  Head is slightly distinct from a moderately slender body. Grows to 0.6 m in length. Dorsally, reddish-brown, with each scale having a fine black edging forming a netted appearance. The head is black on top, divided by a yellow-brown stripe across the snout. The lips and ventral surface are white. Midbody scales are in 15 rows, ventrals number 140–180, subcaudals 25–40 (single) and the anal scale is single.

**Habits and habitat**  Nocturnal. Found in heathlands, shrublands and dry sclerophyll forests, where it seeks refuge under rocks, loose bark at the base of trees and logs. Several specimens may occupy the same site. It is a live bearer and produces 3–7 young. Feeds on small lizards and smaller snakes. Nervous in disposition and may bite if provoked.

# Hooded or monk snake

*Unechis (Rhinoplocephalus) monachus* (Storr, 1964)
*(Venomous)*

**Description** Head is distinct from a moderately slender body. Grows to 0.53 m in length. Dorsally, bright brick red to olive-brown, fading to a lighter colour on the sides adjacent to the ventrals. The head and nape are glossy dark brown or black. The lips and ventral surface are white. Midbody scales are in 15 rows, ventrals number 140–175, subcaudals 21–33 (single) and the anal scale is single.

**Habits and habitat** Nocturnal and seeks refuge under rocks, logs and other debris in *Acacia* woodlands, shrublands and rocky outcrops. It is a live bearer. Probably feeds on small lizards.

# Black-striped snake

*Unechis (Rhinoplocephalus) nigrostriatus* (Krefft, 1864)
*(Venomous)*

**Description** Head is barely distinct from a moderately slender body. Grows to 0.6 m in length. Dorsally, pink to reddish-brown, fading on the lower flanks, with a dark vertebral stripe extending from the head to the tip of the tail. The head is dark brown to black, with white. The lips, neck and ventral surface are whitish. Midbody scales are in 15 rows, ventrals number 160–190, subcaudals 45–70 (single) and the anal scale is single.

**Habits and habitat** Nocturnal. Occurs in dry sclerophyll forests and woodlands, where it seeks refuge under logs, rocks and dead vegetation. It is a live bearer. Feeds on geckoes and other small lizards.

# Bandy bandy

*Vermicella annulata annulata* (Gray, 1841)
*(Venomous)*

**Description** Head is indistinct from a moderately robust body and a short blunt tail. The eyes are small. Grows to 0.75 m in length. Dorsally, white with approximately 30 bluey-black rings. The head is dark and divided by a white bar across the snout, which folds over the front of the head and under the eyes. The ventral surface is similar to the dorsal colour. Midbody scales are in 15 rows, ventrals number 180–243, subcaudals 12–30 (divided) and the anal scale is divided.

**Habits and habitat** A nocturnal fossorial snake, found in a variety of habitats from deserts to rainforests. Found under rocks, in cracks in the soil, termite mounds and logs. Often active at night after rain. Oviparous and produces 2–13 eggs. Evidence suggests males are sexually mature after 24 months and females after 36 months. Females attain a much larger body size than males. Feeds mainly on blind snakes, but will take small lizards. Its feeding frequency is very small (only about one-twentieth of other Australian elapids). Food items are large and can be larger than the *Vermicella* feeding on them. Its defence posture is spectacular, as it flattens its body and forms a number of elevated loops, displaying its contrasting colours. A theory of 'flicker fusion', which allows this snake to escape whilst appearing to be stationary in dim light, has also been offered to explain the banding on this species.

## Subspecies

### *Vermicella annulata snelli* (Storr, 1968)

Differs from *V. a. annulata* in having more ventral scales (254–313) and usually a greater number of rings (34–66). Habits and habitat are similar to *V. a. annulata*.

# Northern bandy bandy

*Vermicella multifasciata* (Longman, 1915)
*(Venomous)*

**Description**  Small head is indistinct from a slender body. Grows to 0.54 m in length. Dorsally, black with 46–83 narrow white rings. The head is black with a white bar across the snout and a white flare around the neck, but not connecting in the middle. The ventral surface is similar to dorsal colour, but paler. Midbody scales are in 15 rows, ventrals number 254–282, subcaudals 18–24 (divided) and the anal scale is divided.

**Habits and habitat**  A nocturnal burrowing snake found in all available habitats within its range. Shelters under rocks, in cracks in the soil, termite mounds and logs. Often active at night after rain. It is oviparous. Feeds on blind snakes, but will take small lizards. Its defence posture is spectacular, as it flattens its body and forms a number of elevated hoops, displaying its contrasting colours.

# Sea snakes

## *Families HYDROPHIIDAE and LATICAUDIDAE*

**Hydrophiidae** — Closely allied to terrestrial snakes. They have valvular nostrils, a lingual fossa, a vertically compressed, paddle-like tail and higher muscular and membranous development to the saccular lung, which is the posterior extension of the lung used for air storage. They all possess small fangs, and some are either equally or up to four times as deadly as the Indian cobra *Naja naja*. The venom of some species has not been tested yet, so as a precaution all species of sea snake should be regarded as dangerous. They possess a salt gland or sublingual gland, which is used for salt excretion. The inward and outward movement of the tongue removes the concentrated brine from the gland. They are all viviparous (live bearers) and the young are born underwater.

**Laticaudidae** — Closely allied to terrestrial snakes. Referred to as sea kraits, they are typified by numerous dark bands, valvular nostrils, which are laterally placed, broad ventral scales and a vertically compressed paddle-like tail. Often found in large numbers on land, where they occupy crevices and rocky shorelines. The venom of both Australian species is potent and more toxic than the Indian cobra *Naja naja*.

## *Acalyptophis peronii* (Duméril, 1853)
### *(Highly venomous)*

**Description** Has valvular nostrils, a notch in the rostral scale to allow protrusion of the tongue when the mouth is closed and a vertically compressed, paddle-like tail. Juveniles have tubercles and adults have conspicuous projecting spines on the supraocular or postocular scales. Head is pale and indistinct from a moderately robust body, which is thicker in the middle. Grows to 1.2 m in length. Dorsally, cream, grey or light brown, with black crossbands, which are widest on the mid-vertebral line. Banding is more pronounced in juveniles and less pronounced in older specimens. The lighter interspaces also have scattered black scales. The bands on the tail are less distinct. The scales are keeled (the keel being short and centrally located in each scale) and are imbricate anteriorly. Midbody scales are in 23–31 rows, ventrals are about the same size as the body and number 140–210, and the anal scale is divided.

**Habits and habitat** Lives on coral reefs and feeds on fish. Often seen on the surface, especially at night. It is live bearing, and a small number of records so far indicate it produces 8–10 young.

## *Aipysurus apraefrontalis* (Smith, 1926)
### *(Highly venomous)*

**Description**  Has valvular nostrils, a notch in the rostral scale to allow protrusion of the tongue when the mouth is closed and a vertically compressed, paddle-like tail. Juveniles have tubercles and adults have conspicuous projecting spines on the supraocular or postocular scales. Head is small and dark, indistinct from a moderately robust body. Grows to 1.0 m in length. Dorsally, creamish to purplish-brown to dark olive-brown, with numerous crossbands, which are more conspicuous on the lower sides and sometimes only faintly defined. The throat has small scales, which are creamy-white anteriorly and dark brown posteriorly. The ventral surface is more or less uniformly grey-brown, with scattered lighter scales. The body scales are smooth or slightly keeled posteriorly, imbricate anteriorly and the hind edge lanceolate or bifid. Midbody scales are in 17 rows, ventrals are about three times the size of the adjacent body scales and number 140–160, subcaudals 18–25 (single) and the anal scale is divided.

**Habits and habitat**  Lives on coral reefs in holes on the reef flats, under dead coral or along the shallow edges. Feeds on fish. Often seen on the surface, especially at night. It is live bearing, and a small number of records so far indicate it produces 8–10 young.

# Dubois's sea snake

*Aipysurus duboisii* (Bavay, 1869)
*(Highly venomous)*

**Description** Has valvular nostrils, a notch in the rostral scale to allow protrusion of the tongue when the mouth is closed and a vertically compressed, paddle-like tail. Juveniles have tubercles and adults have conspicuous projecting spines on the supraocular or postocular scales. Head is slightly distinct from a moderately robust body. Grows to 1.0 m in length. Dorsally, dark purplish-brown or grey, with numerous darker scales arranged in groups forming crossbands, which are wider along the vertebral line and to some degree form a continuous dark line along the vertebral line. The darker markings are less distinct on the tail. The head is intermediate between the dark scales and the lighter ones. Alternatively some specimens are almost black with white markings, forming bands that are discontinuous in the middle. The scales on the throat are white tipped with brown. The ventral surface is cream to dark brown. The snout is pointed. The scales are imbricate and smooth or with a sharp central keel or a series of tubercles. Midbody scales are in 19 rows, ventrals number 150–165 and are slightly notched behind, subcaudals 23–35 (single), and the anal scale is divided.

**Habits and habitat** Lives on coral reefs, venturing down to depths of 50 m. Feeds on fish and eels. It is live bearing.

# Spine-tailed sea snake

*Aipysurus eydouxii* (Gray, 1849)
*(Venomous)*

**Description**  Has valvular nostrils, a notch in the rostral scale to allow protrusion of the tongue when the mouth is closed and a vertically compressed, paddle-like tail. Juveniles have tubercles and adults have conspicuous projecting spines on the supraocular in postocular scales. Head is indistinct from a moderately robust body. Grows to 1.0 m in length. Dorsally, creamish or salmon pink, with irregular black crossbands, which become less intense and may taper laterally or are sometimes absent adjacent to the ventral surface and less distinct on the tail. The lighter scales in the interspaces may have dark margins and sometimes faint secondary bands. The head is light brownish and lighter around the lips. The ventrals are black to pale cream, with a darker medial line. The body scales are smooth and imbricate. Midbody scales are in 17 rows, ventrals are slightly notched behind and number 124–150, subcaudals 23–35 (single) and the anal scale is divided.

**Habits and habitat**  Lives in deeper, more turbid waters (30–50 m) and not the shallow reef areas. Believed to feed on fish eggs. It is live bearing. Its venom is only about one-fortieth as toxic as the Indian cobra *Naja naja*.

## *Aipysurus foliosquama* (Smith, 1926)
### *(Venomous)*

**Description** Has valvular nostrils, a notch in the rostral scale to allow protrusion of the tongue when the mouth is closed and a vertically compressed, paddle-like tail. Juveniles have tubercles and adults have conspicuous projecting spines on the supraocular or postocular scales. Head is pointed and indistinct from a moderately slender body. Grows to 1.0 m in length. Dorsally, brown to black or purplish-brown, with numerous faint crossbands, which are more prominent on the flanks, and scattered paler scales. The ventral surface is grey-brown, with some lighter-coloured scales. The scales are smooth or slightly keeled, imbricate, the hind edge of each scale pointed or bifid. Midbody scales are in 19 or 21 rows, ventrals are large, deeply notched and number 135–155, subcaudals 20–30 (single), and the anal scale is divided.

**Habits and habitat** Lives in shallow waters (less than 10 m) and feeds on fish. It is live bearing.

## *Aipysurus fuscus* (Tschudi, 1837)
### *(Highly venomous)*

**Description** Has valvular nostrils, a notch in the rostral scale to allow protrusion of the tongue when the mouth is closed and a vertically compressed, paddle-like tail. Juveniles have tubercles and adults have conspicuous projecting spines on the supraocular or postocular scales. Head is indistinct from a moderately slender body, with a pointed snout, but blunter than *A. foliosquama*. Grows to 1.0 m in length. Dorsally, brown to blackish, with faint pale cross-bands and thinner dark striations longitudinally on the flanks. The ventral surface is brown. The scales are smooth and imbricate. Mid-body scales are in 19 (rarely 21) rows, ventrals are slightly notched behind and number 155–180, subcaudals 20–40 (single), and the anal scale is single.

**Habits and habitat** Lives on coral reefs in water less than 10 m. Feeds on fish and their eggs. It is live bearing.

# Olive sea snake

*Aipysurus laevis* (Lacépède, 1804)
*(Highly venomous)*

**Description** Has valvular nostrils, a notch in the rostral scale to allow protrusion of the tongue when the mouth is closed and a vertically compressed, paddle-like tail. Juveniles have tubercles and adults have conspicuous projecting spines on the supraocular or postocular scales. Head is slightly distinct from a robust body. Grows to 2.0 m in length. Dorsally, dark brown, purplish-brown to creamy-yellow, with scattered lighter scales. The darker scales often have darker centres, which form longitudinal striations. The tail is uniform brown to pure white, with a dark brown dorsal ridge. The lower flanks and ventrals may be creamy-white or brown. The body scales are smooth and imbricate. Midbody, scales are in 21–25 rows, ventrals number 25–35 (single) and the anal scale is divided.

**Habits and habitat** Lives on coral reefs and feeds on fish, shrimps and molluscs. It is live bearing, and 2–5 large young have been recorded. One of the most abundant snakes on the coral reefs and is reported to be most curious around divers. Its venom is slightly more toxic than the Indian cobra *Naja naja* and is known to be predominantly neurotoxic. Captive observations have shown it uses constriction to restrain prey whilst the venom immobilises it.

# *Aipysurus tenuis* (Lönnberg & Anderson, 1913)
## *(Highly venomous)*

**Description** Has valvular nostrils, a notch in the rostral scale to allow protrusion of the tongue when the mouth is closed and a vertically compressed, paddle-like tail. Juveniles have tubercles and adults have conspicuous projecting spines on the supraocular or postocular scales. Head is indistinct from a moderately robust body. Grows to 1.02 m in length. Dorsally, light brown with dark brown tips on the scales, which form longitudinal lines along the medial line and distinct crossbands on the sides. The head is darker. The scales are smooth (or the outer rows have small tubercles) and imbricate. Midbody scales are in 19 rows, ventrals have a small notch behind, a small median keel and a series of tubercles along the hind edge, and number 185–195, subcaudals 35–40 (single) and the anal scale is divided.

**Habits and habitat** Habits are unknown. It is live bearing.

# Stokes's sea snake

*Astrotia stokesii* (Gray, 1846)
*(Highly venomous)*

**Description**  Has valvular nostrils, a notch in the rostral scale to allow protrusion of the tongue when the mouth is closed and a vertically compressed, paddle-like tail. Juveniles have tubercles and adults have conspicuous projecting spines on the supraocular or postocular scales. Head is distinct from a robust body. Grows to 1.5 m in length. Dorsally, cream, grey to almost black, with darker crossbands, which only extend halfway down each side. Banding in juveniles is more pronounced. The head shields are large and regular. The scales are keeled. The body scales are imbricate, with a median keel or a series of tubercles. Midbody scales are in 45–63 rows, ventrals (except for those on the throat which are single) are divided into pairs of foliform scales, which are about the same size as adjacent body scales and form a deep conspicuous keel (not in juveniles). They number 226–286 and the preanal scales are enlarged.

**Habits and habitat**  Usually seen on the surface of turbid coastal and reef waters (less than 10 m), where it feeds on fish (especially frogfish). Often found in extensive chains (up to 96 km long) of presumably millions of specimens in the open sea. It is live bearing and one record reports 5 large young. When disturbed, will not hesitate to bite. It is only about 80 per cent as deadly as the Indian cobra *Naja naja*. Venom is known to be neurotoxic.

## *Disteira kingii* (Boulenger, 1896)
### *(Highly venomous)*

**Description** Has valvular nostrils, a notch in the rostral scale to allow protrusion of the tongue when the mouth is closed and a vertically compressed, paddle-like tail. Juveniles have tubercles and adults have conspicuous projecting spines on the supraocular or postocular scales. Head is small and indistinct from a slender body. Grows to 1.5 m in length. Dorsally, greyish with 45–50 darker cross-bands, which only extend about halfway down each side. The head and throat are black, and there are yellowish-white rings around the eyes. The scales are keeled and imbricate. Midbody scales number 37–39, ventrals are noticeably larger than adjacent body scales and number 299–342, and the anal scales are not enlarged.

**Habits and habitat** Normally found in deeper waters and does not appear to be a reef dweller. It is live bearing.

## *Disteira major* (Shaw, 1802)
### *(Highly venomous)*

**Description** Has valvular nostrils, a notch in the rostral scale to allow protrusion of the tongue when the mouth is closed and a vertically compressed, paddle-like tail. Juveniles have tubercles and adults have conspicuous projecting spines on the supraocular or postocular scales. Head is slightly distinct from a moderately robust body. Grows to 1.5 m in length. Dorsally, light grey with 25–30 dark crossbands, which do not extend to the ventral surface. In the interspaces between the crossbands, there are narrower secondary bands. Below these, and between the ventral surface and the termination of the band, a large lateral blotch is located. Below these, and in line with the wider bands, a series of smaller blotches are located. The head is light brown or olive. Most ventral scales are dark grey tipped. The scales are imbricate and mostly with a low blunt keel. Midbody scales are in 37–43 rows, ventrals are only slightly larger than adjacent body scales and number 230–266, and the anal scales are enlarged.

**Habits and habitat** Usually found in deeper turbid waters and feeds on fish. It is live bearing. Its venom is about 1.5 times more toxic than the Indian cobra *Naja naja*.

## *Emydocephalus annulatus* (Krefft, 1869)
### *(Highly venomous)*

**Description**  Has valvular nostrils, a notch in the rostral scale to allow protrusion of the tongue when the mouth is closed and a vertically compressed, paddle-like tail. Juveniles have tubercles and adults have conspicuous projecting spines on the supraocular or postocular scales. Head is small and indistinct from a moderately robust body. Grows to 1.0 m in length. Dorsally, black, brown and dark grey above and below, with or without numerous creamish crossbands with scattered lighter or darker scales. The head may be creamy-white or yellow above, speckled and blotched with dark brown, with a dark brown band extending from behind the eye to the first band on the nape. The scales are smooth and imbricate. Midbody scales are in 15–17 rows, ventrals have small tubercles, a median keel and number 125–145, subcaudals 20–40 (single) and the anal scale is usually single.

**Habits and habitat**  Lives on coral reefs in large numbers, in waters less than 30 m. Feeds on fish eggs. It is live bearing.

# Beaked sea snake

*Enhydrina (Disteira) schistosa* (Daudin, 1803)
*(Highly venomous)*

**Description** Has valvular nostrils, a notch in the rostral scale to allow protrusion of the tongue when the mouth is closed and a vertically compressed, paddle-like tail. Juveniles have tubercles and adults have conspicuous projecting spines on the supraocular or postocular scales. Head is distinct from a moderately robust body. Grows to 1.5 m in length. Dorsally, greyish with numerous darker crossbands, which only extend about halfway down the flanks. The head is dark grey above and a lighter grey below. The scales are imbricate and shortly keeled. Midbody scales are in 49–66 rows, ventrals are only slightly wider than adjacent body scales and number 239–332.

**Habits and habitat** Lives mainly in coastal estuaries and feeds on fish. Has been recorded taking catfish. It is live bearing. Its venom is potently neurotoxic and muscle destroying, and is 4 times more toxic than the Indian cobra *Naja naja*. Can become aggressive if annoyed.

*Ephalophis greyi* (Smith, 1931)
*(Highly venomous)*

**Description**  Has valvular nostrils, a notch in the rostral scale to allow protrusion of the tongue when the mouth is closed and a vertically compressed, paddle-like tail. Juveniles have tubercles and adults have conspicuous projecting spines on the supraocular or postocular scales. Head is indistinct from a moderately robust body. Grows to 0.66 m in length. Dorsally, creamish to pale olive-brown, with numerous dark blotches or crossbands, which are wider on the vertebral line and sometimes fused with adjacent bands. The head is mottled with dark grey. The scales are smooth, keeled (at least posteriorly) and imbricate. Midbody scales are in 19–21 rows, ventrals are nearly as broad as the body and number 159–169, subcaudals 28–33 (single) and the anal scale is divided.

**Habits and habitat**  Prefers estuarine environments. It is live bearing.

*Hydrelaps darwiniensis* (Boulenger, 1896)
*(Highly venomous)*

**Description** Has valvular nostrils, a notch in the rostral scale to allow protrusion of the tongue when the mouth is closed and a vertically compressed, paddle-like tail. Juveniles have tubercles and adults have conspicuous projecting spines on the supraocular or postocular scales. Head is slightly distinct from a moderately robust body. Grows to 0.5 m in length. Dorsally, cream or yellowish, with 30–45 dark crossbands (about twice as wide as the lighter interspaces) extending the full length of the body, including the tail and complete on the ventral surface. The head is dark and mottled with yellow. The body scales are smooth and imbricate. Midbody scales are in 25–29 rows, ventrals are narrow and about 3 times as wide as the adjacent body scales, numbering 163–172, subcaudals 27–39 (mainly single, but occasionally a few divided anteriorly) and the anal scale is divided.

**Habits and habitat** Often found on mud flats associated with mangroves, where it feeds on small fish. It is live bearing. Timid and inoffensive by nature.

*Hydrophis atriceps* (Günther, 1864)
*(Highly venomous)*

**Description** Has valvular nostrils, a notch in the rostral scale to allow protrusion of the tongue when the mouth is closed and a vertically compressed, paddle-like tail. Juveniles have tubercles and adults have conspicuous projecting spines on the supraocular or postocular scales. Head is indistinct from a slender body anteriorly and deep and compressed posteriorly. Grows to 1.1 m in length. Dorsally, light brown to creamish, with numerous blotches, cross-bands or rings. The head is black. The ventrals are blackish anteriorly. The body scales are imbricate and the midbody scales are in 35–49 rows, ventrals are single, slightly wider than adjacent body scales and number 323–514.

**Habits and habitat** Habits and habitat unknown.

*Hydrophis belcheri* (Gray, 1849)
*(Highly venomous)*

**Description** Has valvular nostrils, a notch in the rostral scale to allow protrusion of the tongue when the mouth is closed and a vertically compressed, paddle-like tail. Juveniles have tubercles and adults have conspicuous projecting spines on the supraocular or postocular scales. Head is indistinct from a moderately robust body. Grows to 1.0 m in length. Dorsally, greyish with about 60 dark crossbands, which are wider on the vertebral line and less distinct on the ventral surface. The ventral surface is pale grey. The head is dark. Midbody scales are in 34 rows, ventrals are mostly single (especially posteriorly) and number 300–320.

**Habits and habitat** Habits and habitat unknown. The only known specimen is from New Guinea and it is assumed it occurs in Australian waters.

*Hydrophis caerulescens* (Shaw, 1802)
*(Highly venomous)*

**Description** Has valvular nostrils, a notch in the rostral scale to allow protrusion of the tongue when the mouth is closed and a vertically compressed, paddle-like tail. Juveniles have tubercles and adults have conspicuous projecting spines on the supraocular or postocular scales. Head is indistinct from a moderately robust body. Grows to 0.9 m in length. Dorsally, bluish-grey with numerous black crossbands, which may be indistinct in some specimens, divided or alternating about the vertebral line on the full length of the body and tail. The body scales are imbricate. Midbody scales are in 37–39 rows, ventrals are slightly wider than adjacent body scales, are mostly single (except anteriorly) and number 266–287.

**Habits and habitat** Habits and habitat unknown. It is live bearing.

## *Hydrophis coggeri* (Kharin, 1984)
### *(Highly venomous)*

**Description** Has valvular nostrils, a notch in the rostral scale to allow protrusion of the tongue when the mouth is closed and a vertically compressed, paddle-like tail. Juveniles have tubercles and adults have conspicuous projecting spines on the supraocular or postocular scales. Head is small and indistinct from a slender body anteriorly and compressed posteriorly. Grows to 1.2 m in length. Dorsally, yellowish to olive-grey, with 30–40 black crossbands on the body and tail. The head is yellow or olive-grey, with distinctive black mottling. The body scales are imbricate and at the midbody are in 29–34 rows, ventrals are small, mostly single and number 278–317.

**Habits and habitat** Prefers deeper waters of about 30–40 m. Occurs in the Coral Sea. Preys on burrowing eels. It is live bearing.

*Hydrophis czeblukovi* (Kharin, 1984)
*(Highly venomous)*

**Description** Has valvular nostrils, a notch in the rostral scale to allow protrusion of the tongue when the mouth is closed and a vertically compressed, paddle-like tail. Juveniles have tubercles and adults have conspicuous projecting spines on the supraocular or postocular scales. Dorsally, grey with 40 black dorsal blotches and a less distinct series of lateral blotches in the pale interspaces. The body scales are keeled and at the midbody are in 55–56 rows with 2–4 tubercles or spines.

**Habits and habitat** Nothing is known of habits or habitat. It is a live bearer.

## *Hydrophis elegans* (Gray, 1842)
### *(Highly venomous)*

**Description** Has valvular nostrils, a notch in the rostral scale to allow protrusion of the tongue when the mouth is closed and a vertically compressed, paddle-like tail. Juveniles have tubercles and adults have conspicuous projecting spines on the supraocular or postocular scales. Head is indistinct from a slender body anteriorly and compressed posteriorly. Grows to 2.0 m in length. Dorsally, highly variable in colour, from pale grey to brown, with numerous dark crossbands or blotches and darker splotches in the paler inter-spaces. Juveniles are more vividly marked. The head is grey, olive-yellow or black. The body scales are imbricate. Midbody scales are in 37–49 rows, ventrals are only slightly wider than adjacent body scales (except anteriorly), are single and number 345–432.

**Habits and habitat** An abundant species found in turbid reef waters greater than 30 m. Preys on eels. It is live bearing. The venom is about 1.1 times more deadly than the Indian cobra *Naja naja*.

## *Hydrophis gracilis* (Shaw, 1802)
## *(Highly venomous)*

**Description** Has valvular nostrils, a notch in the rostral scale to allow protrusion of the tongue when the mouth is closed and a vertically compressed, paddle-like tail. Juveniles have tubercles and adults have conspicuous projecting spines on the supraocular or postocular scales. Head is small and indistinct from a slender body anteriorly and strongly compressed posteriorly. Grows to 1.0 m in length. Dorsally, black with a series of whitish markings, which are usually complete bands posteriorly and are spots anteriorly, numbering 40–60 in total. The ventral surface is whitish. The tubercles or keels on the lower lateral scales are often dark brown or black. The body scales are juxtaposed and the midbody scales are in 30–36 rows, the lower lateral scales are larger than the dorsal scales, ventrals number 220–287.

**Habits and habitat** Forages on sandy bottoms and feeds on fish eggs. It is live bearing.

## *Hydrophis inornatus* (Gray, 1849)
### *(Highly venomous)*

**Description**  Has valvular nostrils, a notch in the rostral scale to allow protrusion of the tongue when the mouth is closed and a vertically compressed, paddle-like tail. Juveniles have tubercles and adults have conspicuous projecting spines on the supraocular or postocular scales. Head is barely distinct from a moderately robust body. Grows to 0.9 m in length. Dorsally, bluey-grey with numerous dark blotches or rings, which are less distinct or reduced to dorsal blotches when adults. The lower half of the body is whitish. The head shields are large and regular. The body scales are imbricate. Midbody scales are in 35–48 rows, ventrals are slightly wider than adjacent body scales, mostly single and number 195–293.

**Habits and habitat**  Only known from a few specimens from Australian waters. It is live bearing.

*Hydrophis mcdowelli* (Kharin, 1983)
*(Highly venomous)*

**Description** Has valvular nostrils, a notch in the rostral scale to allow protrusion of the tongue when the mouth is closed and a vertically compressed, paddle-like tail. Juveniles have tubercles and adults have conspicuous projecting spines on the supraocular or postocular scales. The very small head is indistinct from a slender body anteriorly and strongly compressed posteriorly. Grows to 1.0 m in length. Dorsally, creamish with numerous dark blotches or bands, which tend to be complete rings anteriorly but incomplete posteriorly, extending laterally only about a third of the way down the body. Other blotches occur on the flanks. The head is dark olive or black. The ventral surface is pale yellowish. The throat and anterior ventrals are grey to black. The scales are keeled. The body scales are imbricate. Midbody scales are in 35–39 rows, ventrals are mostly single and number 252–274.

**Habits and habitat** Found both in deeper waters and coastal estuaries. It is live bearing and 2–3 large young have been recorded.

## *Hydrophis melanosoma* (Günther, 1864)
## *(Highly venomous)*

**Description**  Has valvular nostrils, a notch in the rostral scale to allow protrusion of the tongue when the mouth is closed and a vertically compressed, paddle-like tail. Juveniles have tubercles and adults have conspicuous projecting spines on the supraocular or postocular scales. Head is indistinct from a moderately robust body. Grows to 1.5 m in length. Dorsally, creamish or yellowish, with 50–70 broad black crossbands. The head is black. The body scales are imbricate. Midbody scales are in 37–43 rows, ventrals are wider than adjacent body scales, are single and number 260–370.

**Habits and habitat**  Abundant in the shoal waters off northern Australia. Is is live bearing.

*Hydrophis ornatus (ocellatus)* (Gray, 1842)
*(Highly venomous)*

**Description** Has valvular nostrils, a notch in the rostral scale to allow protrusion of the tongue when the mouth is closed and a vertically compressed, paddle-like tail. Juveniles have tubercles and adults have conspicuous projecting spines on the supraocular or postocular scales. Head is indistinct from a moderately robust body. Grows to 1.2 m in length. Dorsally, blue-grey above with 30–60 darker crossbands or blotches and other larger and smaller blotches on the lateral surfaces. Markings are less distinct with age. The head is grey and paler around the lips. The lower half of the body, including the ventral surface, is pale cream. The body scales are imbricate. Midbody scales are in 39–59 rows, ventrals are about twice the size of adjacent body scales and number 240–340.

**Habits and habitat** Widely distributed throughout the coastal waters of northern Australia and has been recorded as far south as Tasmania. Conflicting reports over its preference of water depth exist. Some claim it prefers deeper waters of greater than 30 m and others say it prefers shallow water. It is live bearing. Its venon is about 1.8 times more deadly than the Indian cobra *Naja naja*.

*Hydrophis pacificus* (Boulenger, 1896)
*(Highly venomous)*

**Description** Has valvular nostrils, a notch in the rostral scale to allow protrusion of the tongue when the mouth is closed and a vertically compressed, paddle-like tail. Juveniles have tubercles and adults have conspicuous projecting spines on the supraocular or postocular scales. The head is slightly distinct from a slender body anteriorly, but deep and compressed posteriorly. Grows to 1.5 m in length. Dorsally, dark grey and whitish ventrally, the two zones meeting at a distinct midlateral zone. Numerous dark bands over the whole body, joining dark ventrals anteriorly, often expanded along the vertebral line. Juveniles are more vividly marked. The head is dark, speckled with grey. Body scales are imbricate. Mid-body scales are in 45–49 rows, ventrals are not much wider than adjacent scales on the body, are mostly single and number 320–430.

**Habits and habitat** Its habits are largely unknown. It is live bearing.

*Hydrophis vorisi* (Kharin, 1984)
*(Highly venomous)*

**Description** Has valvular nostrils, a notch in the rostral scale to allow protrusion of the tongue when the mouth is closed and a vertically compressed, paddle-like tail. Juveniles have tubercles and adults have conspicuous projecting spines on the supraocular or postocular scales. The small head is indistinct from a slender body anteriorly and compressed posteriorly. Grows to 1.2 m in length. Dorsally, white with numerous dark crossbands circling the body and expanded along the dorsal line. Body scales are imbricate. Mid-body scales are in 29–35 rows, ventrals are hardly larger than adjacent scales on the body (except anteriorly), are mostly single and number 330–350.

**Habits and habitat** Only a small number of specimens have been recorded from northern Australia. It is live bearing.

*Lapemis hardwickii* (Gray, 1835)
*(Highly venomous)*

**Description** Has valvular nostrils, a notch in the rostral scale to allow protrusion of the tongue when the mouth is closed and a vertically compressed, paddle-like tail. Juveniles have tubercles and adults have conspicuous projecting spines on the supraocular or postocular scales. Head is indistinct from a moderately robust body. Grows to 1.2 m in length. Dorsally, pale olive, greenish or grey, merging into cream or pale yellow below. A series of darker bands (30–55) may be present and are more conspicuous in juveniles. These tend to be joined along the mid-dorsal line. The head is either same colour as dorsal markings or brown. Body scales are juxtaposed and hexagonal, and the lower lateral scales are much larger, with grossly enlarged tubercles in adult males. These spinous scales are thought to serve as a fluid balance between the body of the snake and the external sea water, as well as performing some sensory function. Midbody scales are in 23–46 rows, ventrals are about the same size as adjacent body scales, are unrecognisable posteriorly and number 110–240; the preanal scales are scarcely enlarged.

**Habits and habitat** Widely distributed in northern waters and found in a large variety of habitats, from coral reefs to estuaries. Feeds on fish and is live bearing. Its venom is about half as toxic as the Indian cobra *Naja naja*, and is known to be neurotoxic.

## *Parahydrophis mertoni* (Roux, 1910)
### *(Highly venomous)*

**Description** Has valvular nostrils, a notch in the rostral scale to allow protrusion of the tongue when the mouth is closed and a vertically compressed, paddle-like tail. Juveniles have tubercles and adults have conspicuous projecting spines on the supraocular or postocular scales. Head is indistinct from a moderately robust body. Grows to 0.6 m in length. Dorsally, blue-grey, grey-brown or yellow, with numerous dark crossbands often fused in the mid-dorsal line. The head is brown with black mottling. The snout and side of the head are blackish. The scales are smooth and imbricate. Midbody scales are in 36–39 rows, ventrals are large (about 3 times as wide as adjacent body scales) and number 158–160, subcaudals 29–35 (single except for 1 or 2 anterior scales) and the anal scale is divided.

**Habits and habitat** Found mainly in estuaries within its range. Feeds on fish. It is live bearing and one specimen was found to have 3 large developing young.

# Yellow-bellied sea snake

*Pelamis platurus* (Linnaeus, 1766)
*(Highly venomous)*

**Description**  Has valvular nostrils, a notch in the rostral scale to allow protrusion of the tongue when the mouth is closed and a vertically compressed, paddle-like tail. Juveniles have tubercles and adults have conspicuous projecting spines on the supraocular or postocular scales. Head is indistinct from a moderately robust body. Grows to 0.8 m in length. Dorsally, black or dark brown from the head to the tail, contrasting with a bright yellow, cream or light brown on lower half. These two colours meet at a distinct mid-lateral line. The tail is yellow with black bands or spots. The body scales are juxtaposed, hexagonal or triangular. Midbody scales are in 47–69 rows, ventrals are divided, scarcely wider than adjacent body scales and number 264–406.

**Habits and habitat**  Found along much of the Australian coast-line, but not the southern section. Seems to prefer the vast water mass rather than the confines of the coastal waters, and is often found floating on the surface of the water, sometimes in huge slicks or masses of snakes. Feeds on fish. It is live bearing and produces 1–6 young. Its gestation period is 5 months. Its venom is approximately as toxic as the Indian cobra *Naja naja* and is known to be neurotoxic.

# Banded sea krait

*Laticauda colubrina* (Schneider, 1799)
*(Highly venomous)*

**Description** Has lateral nostrils and a vertically compressed, paddle-like tail. Head is slightly distinct from a moderately robust body. Grows to 1.4 m in length. Dorsally, blue-grey with 20–65 black bands, which are continuous on the ventral surface. The ventral surface is cream or yellow. The lips, snout and a streak from the eye to the temporal region are yellow; the rest of the head is black. Body scales are imbricate and smooth. Midbody scales are in 21–25 rows, ventrals are about half the width of the body and number 210–250, subcaudals number 25–35 in females and 36–50 in males and are divided, and the anal scale is divided.

**Habits and habitat** Partly terrestrial, having been found some distance from water. It is oviparous and lays its eggs on land. Found along rocky coral crevices, in mangrove swamps and in breakwater constructions, often occurring in large numbers. Mainly nocturnal, but can be found foraging during the day. Feeds on fish and eels. The venom is about 70 per cent as toxic as the Indian cobra *Naja naja* and is known to be neurotoxic.

163

# Banded sea krait

*Laticauda laticauda* (Linnaeus, 1758)
*(Highly venomous)*

**Description** Has lateral nostrils and a vertically compressed, paddle-like tail. Head is slightly distinct from a moderately robust body. Grows to 1.0 m in length. Dorsally, blue-grey with 25–70 black bands, some or all of which may be discontinuous on the ventral surface. The ventral surface is cream or yellow. The supralabials are brown; the snout and a streak from the eye to the temporal region are yellow; the rest of the head is black. Body scales are imbricate and smooth. Midbody scales are in 19 rows, ventrals are about half the width of the body and number 225–245, subcaudals number 25–35 in females and 36–50 in males and are divided, and the anal scale is divided.

**Habits and habitat** Partly terrestrial, having been found some distance from water. It is oviparous and lays its eggs on land. Found along rocky coral crevices, in mangrove swamps and in breakwater constructions, often occurring in large numbers. Mainly nocturnal, but can be found foraging during the day. Feeds on fish and eels. The venom is about 1.7 times as toxic as the Indian cobra *Naja naja* and is known to be neurotoxic.

# Solid-toothed and rear-fanged snakes

## *Family COLUBRIDAE*

Although widespread in other parts of the world, this family is only poorly represented in Australia. They occur mainly in the northern part of Australia. The tail is more or less cylindrical; they have a single row of enlarged ventral scales, fewer than 30 midbodies; the presence of a loreal shield or if absent, 23 or more midbody scales and generally a divided anal scale.

There are two subfamilies:

**Colubrinae** — Solid-toothed species represented by *Dendrelaphis*, *Styporhynchus*, and *Stegonotus* and the weakly venomous, rear-fanged represented by *Boiga*.

**Homalopsinae** — Rear-fanged and weakly venomous, aquatic snakes represented by *Cerberus*, *Fordonia*, *Myron* and *Enhydris*.

# Brown tree snake, night tiger or doll's eye snake

*Boiga irregularis* (Merrem, 1802)
*(Venomous, rear-fanged)*

**Description**  A large bulbous head with protruding eyes and vertical pupils distinct from a long slender body. Grows to 2.0 m in length. Dorsally, white with numerous red, brown or black crossbands on the body and tail or it may be cream or brown with numerous darker crossbands on the body and tail. In some cases, the crossbands may be barely visible. The ventral surface is cream or salmon coloured. Midbody scales are in 19–23 rows, ventrals number 225–265, subcaudals 85–130 (divided) and the anal scale is single.

**Habits and habitat**  Nocturnal and arboreal, but sometimes forages on the ground. Found in wet and dry sclerophyll forests, rainforests, woodlands, rocky outcrops, vine thickets and rocky gorges, where it seeks refuge in hollow trees, caves, termite mounds, hollow logs, rock crevices, under stones and rocks. If provoked, it puts on a brave defensive display by placing its forebody in an S-shaped posture, hissing and lunging with its mouth agape at nearby movement. Fortunately, it is not dangerous to humans, and its striking speed is relatively slow when compared with the elapids. There have been no cases of envenomation by this species despite the many bites resulting from captive specimens in the charge of herpetologists. It feeds on mammals, birds and their eggs, and lizards, and has even been recorded using constriction as method of prey restraint. It is an egg layer, producing 4–12 eggs in a clutch.

# Bockadam

*Cerberus rynchops* (Schneider, 1799)
*(Venomous, rear-fanged)*

**Description** Broad head is distinct from a moderately robust body. Valvular nostrils on top of snout. The eyes are small, with vertical pupils, directed upwards and protruding, with raised supra-ocular scales. Grows to 1.2 m in length. Dorsally, grey to reddish with numerous black elongated spots orientated transversely, forming broken bands. Dark streaks are present from the snout through the eye and another from the angle of the mouth to the neck. The sides are often flushed with pink. The ventral surface is whitish, cream, yellow or salmon. The body scales are keeled. Midbody scales are in 25 (occasionally 23) rows, ventrals number 140–160, subcaudals 45–60 (divided) and the anal scale is divided.

**Habits and habitat** Nocturnal and amphibious, living near estuaries, tidal creeks, rivers and mangrove flats, where it shelters under mangrove roots or, when the water recedes at low tide, burrows into the mud, leaving just its eyes protruding. Often occurs in small colonies. Feeds on mudskippers, fish and crustaceans. It is a live bearer and produces 1–26 young. Its bite is not considered dangerous, but it can cause local stinging. Can be pugnacious if provoked, biting frequently and emitting a foul smell from the anal gland.

# Northern tree snake

*Dendrelaphis calligaster* (Günther, 1867)
*(Non-venomous)*

**Description** Head is slightly distinct from a long, extremely slender body. It has a large eye and round pupil. Grows to 1.2 m in length. Dorsally, olive, olive-grey to light to rich brown, fading on the flanks. There is a distinct dark streak extending from the snout through the eye to the neck, which contrasts with the yellow, cream-coloured scales of the bottom jaw and lower side of the neck. Juveniles may exhibit some banding. The ventral surface is white to yellow, dotted with black specks. Midbody scales are in 13–15 rows, ventrals number 180–230, subcaudals 90–150 (divided) and the anal scale is divided. Specimens often found with lumps under their skin. These are caused by a parasite skin worm and are typical in snakes feeding predominantly on frogs. No fangs are present.

**Habits and habitat** Mostly diurnal and arboreal, but can be found foraging on the ground. Extremely agile, and can ascend a tree very quickly. Found in rainforests, vine thickets, near water. When it becomes agitated, the skin between the scales becomes visible and its colour is a pale to dark blue. It is an egg layer and produces 5–12 eggs. Feeds mainly on frogs, but will also take lizards.

# Green or common tree snake

*Dendrelaphis punctulatus* (Gray, 1827)
*(Non-venomous)*

**Description** Head is slightly distinct from a long slender body. It has a large eye and a round pupil. Grows to 2.0 m in length. Dorsally, it is variable in colour according to geographical location and may be olive, brown, yellow, blue, green and black, fading on the flanks. In NSW and southern Queensland, they are mostly green dorsally and yellow ventrally. In north Queensland, they are brown to black dorsally and yellow ventrally. There is also an entirely blue form that is widespread but uncommon. Western Australian and Northern Territory groups are yellow with a bluish head. The ventral surface is yellow, green, white, olive and blue. Midbody scales are in 13 (rarely 15) rows, ventrals number 180–230, subcaudals 90–150 (divided) and the anal scale is divided.

**Habits and habitat** Arboreal and diurnal. Found in semi-arid areas to rainforests in a variety of habitats, from woodlands, vine thickets to wet and dry sclerophyll forests, where it shelters in hollow limbs, rock crevices, under debris, among foliage and in caves. In southern areas, large numbers have been found wintering in groups in tree hollows, abandoned houses and caves. It is an egg layer and produces up to 14 eggs. Feeds mainly on frogs and tadpoles, but will also take small lizards. Fish and other snakes (*Hoplocephalus bungaroides*) have also been reported as food items. When disturbed, it inflates its body, displaying the light blue skin between the scales. As a defence mechanism, it can release a foul-smelling odour from its anal gland when molested. This species does not possess fangs.

# Macleay's water snake

*Enhydris polylepis* (Fischer, 1886)
*(Venomous, rear-fanged)*

**Description**   The head is distinct from a moderately robust body. Grows to 1.0 m in length. Eyes are small, set on top of head, with vertical pupils and valvular nostrils. Dorsally, light to dark brown, olive to grey, sometimes with a broad dark vertebral stripe extending from the head, breaking into two parallel stripes anteriorly. Sometimes in Queensland specimens, irregular tranverse bars are present posteriorly. There may be a dark stripe extending from the snout through the eye to the body and breaking towards the tail. The lower sides are yellow or cream, mottled with black. Another colour variation exists in Queensland, and it has been proposed that this is a separate species, *Enhydris macleayi*. It is grey or brown with a light stripe from the snout, under the eye, extending to the neck, and a black streak along the lower sides of the body. Numerous dark bands occur on the posterior two-thirds of the body. Ventral surfaces are cream to yellow, often mottled with black anteriorly and on the lateral edges of the ventral scales. On the underside of the tail there may be a broad black streak, or each subcaudal scale may be black with yellow edging posteriorly. Midbody scales are in 21 or 23 rows, ventrals number 137–165, subcaudals 35–50 (divided) and the anal scale is divided.

**Habits and habitat**   Freshwater aquatic; nocturnal and sometimes diurnal. Found in creeks, swamps, rivers, waterholes and lagoons, where it shelters in submerged roots (especially pandanas), grassy banks, dense aquatic vegetation and caverns. It is widely dispersed in the wet season flooding and contracts to waterholes during the dry season. It is a live bearer and produces up to 15 young. Feeds on aquatic vertebrates, including fish, frogs and tadpoles. It is reported to wait in ambush for its prey, using submerged roots and so on to launch its attack. Although mildly venomous, it is not regarded as dangerous.

# White-bellied mangrove snake

*Fordonia leucobalia* (Schlegel, 1837)
*(Venomous)*

**Description**  It has a rounded snout; the head is distinct from a robust body. The eyes are small with rounded pupils. The nostril is valvular. Grows to 1.0 m in length. Dorsal colouration is extremely variable. May be glossy black, reddish to dark grey, with or without lighter markings. The lighter markings may be white, white edged with black, blue, cream or pale grey, which sometimes form irregular transverse bands. Sometimes the lighter patches predominate over the black. The ventral surface is white or creamish and lower flanks are white, cream or pinkish. Midbody scales are in 23–29 rows, ventrals number 130–160, subcaudals 25–45 (usually divided) and the anal scale is divided.

**Habits and habitat**  Aquatic and nocturnal to diurnal. Found in estuaries, brackish water, mangrove mudflats, tidal creeks and rivers along the coastline, where it shelters in root tangles and crab burrows. It is a live bearer and produces about 8 young in a litter. Feeds on fish and crustaceans (fiddler crabs *Uca* spp.), holding them against any available surface with its body coils, biting off their claws and legs and consuming the body. Although venomous, it is not dangerous. Uses constriction to restrain prey whilst immobilising and devouring it.

# Richardson's mangrove snake

*Myron richardsonii* (Gray, 1849)
*(Venomous)*

**Description** Head is slightly distinct from a relatively slender body. Grows to 0.6 m in length. The eyes are small and slightly protrusive and directed upwards. Nostril is valvular. Dorsal scales are bluntly keeled. Dorsally, brown to olive-grey with numerous darker crossbands, usually extending onto the ventral surface. The head is usually a darker shade of the ground colour. The ventral surface is cream or yellow, with black on the anterior edge to each ventral and subcaudal scale. Midbody scales are in 21–23 rows, ventrals number 130–145, subcaudals 30–40 (divided) and the anal scale is divided.

**Habits and habitat** An aquatic, nocturnal species. Found in the littoral or supralittoral zone of the coastlines, where it inhabits mangroves and mudflats of the tidal estuaries, creeks and rivers. Seeks refuge in mangrove roots, submerged rocks, stones and crab burrows. It is a live bearer and produces about 8 young in a litter. Feeds on small fish and crustaceans. Uses constriction to restrain prey whilst immobilising and devouring it.

# Slaty grey snake

*Stegonotus cucullatus* (Duméril, Bibron & Duméril, 1854)
*(Non-venomous)*

**Description** Broad-snouted head is distinct from a moderately slender body. Grows to 2.0 m in length. Eyes are moderately small, with a rounded pupil and black iris. Dorsal scales are feebly keeled. Dorsally, slaty grey, black to brown, with an iridescent purple sheen in reflected light. The colour fades on the flanks. The lips are pale to cream. The ventral surface is white to cream, sometimes with black flecks on the posterior ventral scales. Midbody scales are in 17 (occasionally 19) rows, ventrals number 170–225, subcaudals 60–105 (divided) and the anal scale is single.

**Habits and habitat** Nocturnal and found close to water. Does not hesitate to enter water. Feeds on frogs, lizards and mammals. Found in adjacent creeks, rivers, swamps and lagoons, woodlands and rainforests, where it seeks refuge under rocks, abandoned burrows and low vegetation. It is an egg layer and produces up to 16 eggs. If provoked, it bites savagely and repeatedly. Does not possess fangs. Uses constriction to restrain prey whilst devouring it.

*Stegonotus parvus* (Meyer, 1874)
*(Non-venomous)*

**Description**  Very similar to *S. cucullatus* in appearance and scalation. Can be distinguished by its paler upper lip and by a wide gap between the third and fourth maxillary teeth. Grows to a smaller size of 0.8 m.

**Habits and habitat**  Very similar habits and habitat to *S. cucullatus*. Australian specimens are only known from Murray Island.

# Freshwater snake or keelback

*Styporynchus (Amphiesma) mairii* (Gray, 1841)
*(Non-venomous)*

**Description** Head is slightly distinct from a moderately robust body. Eyes are moderately large with a round pupil. Grows to 1.0 m in length. Dorsal scales are strongly keeled and may be olive, brown, red, black and grey, with numerous black flecks, which are sometimes arranged in clusters and crossbands. The lips are whitish. The ventral surface is cream, olive, green, brown and salmon. Midbody scales are in 15 (rarely 17) rows, ventrals number 130–165, subcaudals 50–85 (divided) and the anal scale is divided.

**Habits and habitat** Either nocturnal or diurnal, and largely aquatic. Usually found close to fresh water such as lakes, creeks, rivers and lagoons. Its principal food is frogs and tadpoles, but it will take small lizards and fish. It can successfully prey on smaller specimens of the introduced cane toad *Bufo marinus*, its tadpoles and eggs. Seeks refuge in low vegetation, logs and abandoned burrows. It is an egg layer and produces 5–17 eggs. Mating takes place from early October to December; male combat has been observed. Hatching takes place from February to April. The tail is easily shed if trapped. Often confused with the highly venomous *Tropidechis carinatus*. When threatened, it raises its head and forebody, flattens its neck and strikes feebly at any movement. Can emit an unpleasant odour if provoked.

# File snakes
## *Family ACROCHORDIDAE*

These snakes are characterised by robust bodies, rough rasp-like keeled scales, baggy skin and a blunt snout, and no enlarged ventral scales. They have small eyes, which are slightly protrusive, and a vertical pupil. Entirely aquatic, they are nocturnal and diurnal. Their long sharp teeth and rasp-like scales combine to effectively constrict their prey of slippery wet fish. They are live bearing, harmless, with solid teeth and a prehensile tail. They are eaten by Aborigines.

# Arafura file snake

*Acrochordus arafurae* (McDowell, 1979)
*(Non-venomous)*

**Description** A large robust, compressed body and rough tiny finely keeled (file-like) or pointed rasp-like scales on a loose baggy skin. Small protrusive eye, blunt snout and a valvular nostril. The tail is narrow and prehensile. The head is barely distinct from the body. Grows to 2.5 m in length. Dorsally, grey to brownish, with darker variegated markings, which appear to form a reticulum or a series of oblong blotches, often coalescing vertebrally to form a broad line. The head is spotted anteriorly, which tends to merge with the reticulum on the temple and neck. The ventral surface is whitish, with the darker dorsal blotches extending to the ventral surface. Midbody scales are in 120–180 rows. There is no enlarged ventral and subcaudal or anal scales.

**Habits and habitat** Either nocturnal or diurnal and entirely aquatic. Extremely agile in water, but almost helpless on land. Shelters beneath submerged objects, under overhanging root systems and among aquatic vegetation. Able to disperse during the monsoon flooding through ephemeral streams. Found in streams and lagoons, and also estuaries and the sea. Feeds almost entirely on fish. It is live bearing, the largest litter recorded being 27. Many males have been observed in a writhing ball attempting to mate with a single female.

# Little file snake

*Acrochordus granulatus* (Schneider, 1799)
*(Non-venomous)*

**Description** Small and relatively slender, with a compressed robust body and rough tiny finely keeled (file-like) or pointed rasp-like scales on a loose baggy skin, but not as loose as *A. arafurae*. Small protrusive eye, blunt snout and a valvular nostril. The tail is narrow, prehensile and compressed. The head is barely distinct from the body. Grows to 1.2 m in length. Dorsally, grey-brown or almost black, with numerous whitish or fawn bands, which become less distinct on the ventral surface. The head is usually darker with numerous lighter spots. The ventral surface is similar, but is lighter than the dorsal colours. Midbody scales are in 90–160 rows; there is no distinct size difference between the dorsal, ventral, anal or the subcaudal scales. There is a distinct mid-ventral fold.

**Habits and habitat** An aquatic marine and estuarine snake (and occasionally freshwater). Forages among mangroves in the intertidal zone, searching for crabs (gobbies and mudskippers) and fish. Will occasionally bask on the exposed mud flats. It is live bearing, and litters of 2–12 young are recorded, with a gestation period of 5–8 months. Mainly nocturnal and buries itself in the mud or hides under submerged debris during the day.

# Blind snakes
## *Family TYPHLOPIDAE*

Blind or worm snakes are typified by having blunt heads, which are indistinct from the neck. The body scales and ventrals are of similar size, and the tail ends in an abrupt point, which is used for loco-motion. They all have a vestigal pelvis. There are about 30 species currently recognised in Australia, but due to space and relative interest, only 4 species are described in this text.

Blind snakes are fossorial and feed on termites, their eggs, pupae or larvae, and are mostly found in association with termite mounds. They are all considered to be egg layers, although this has not been confirmed with all species. They have the ability to emit an un-pleasant odour from their anal glands. They are usually only seen moving about at night, especially on warm nights after rain.

# *Ramphotyphlops affinis* (Boulenger, 1889)

**Description** A pale head, brownish dorsally, merging to yellowish, cream, pinkish-white on the ventral surface and flanks. Midbody scales are in 18 rows. The snout is rounded when viewed from above, with a bluntly angular profile. Nasal cleft extends from the second upper labial scale to the rostral scale. Grows to about 0.22 m. It has a short tail, ending abruptly in a sharp spine. The eye is greatly reduced and is covered by a scale. The mouth is small and covered by a protrusive snout. The body scales and ventrals are equal in size and are highly polished to provide less friction when the snake is moving through the earth.

**Habits and habitat** Fossorial and shelters in ant nests, termite mounds or termitaria, soil beneath leaf litter, rocks and logs. Rarely active on the ground surface, but most commonly encountered after rains during summer nights. Almost exclusively feeds on termites and their eggs, and the pupae and larvae of ants. Most blind snakes are regarded as egg layers, but this has not been positively determined for this species. Completely harmless and relies on emitting an unpleasant odour from the anal glands to deter predators.

# *Ramphotyphlops australis* (Gray, 1845)

**Description** Moderately robust. Brown, grey-brown or purplish-brown dorsally, merging raggedly to white on the flanks and the ventral surface. Midbody scales are in 22 rows. The snout is bluntly rounded when viewed from above and projecting in profile. Nasal cleft extends from the second upper labial scale to the rostral scale. Grows to about 0.5 m. It has a short tail, ending abruptly in a sharp spine. The eye is greatly reduced and is covered by a scale. The mouth is small and covered by a protrusive snout. The body scales and ventrals are equal in size and are highly polished to provide less friction when the snake is moving through the earth.

**Habits and habitat** Fossorial and shelters in ant nests, termite mounds or termitaria, soil beneath leaf litter, rocks and logs. Rarely active on the ground surface, but most commonly encountered after rains during summer nights. Almost exclusively feeds on termites and their eggs, and the pupae and larvae of ants. Most blind snakes are regarded as egg layers, but this has not been positively determined for this species. Completely harmless and relies on emitting an unpleasant odour from the anal glands to deter predators.

*Ramphotyphlops grypus* (Waite, 1918)

**Description**  Snout pale and the remainder of the head and neck tending to black. Light brown above, creamish below. The tail is usually darker than the remainder of the body. Midbody scales are in 18·rows. The snout is elongate and bluntly angular when viewed from above, with a strongly hooked, recurved beak when viewed in profile. The rostral is elongate when viewed from above. Grows to about 0.45 m. It has a short tail, ending abruptly in a sharp spine. The eye is greatly reduced and is covered by a scale. The mouth is small and covered by a protrusive snout. The body scales and ventrals are equal in size and are highly polished to provide less friction when the snake is moving through the earth.

**Habits and habitat**  Fossorial and shelters in ant nests, termite mounds or termitaria, soil beneath leaf litter, rocks and logs. Rarely active on the ground surface, but most commonly encountered after rains during summer nights. Almost exclusively feeds on termites and their eggs, and the pupae and larvae of ants (especially red meat ants). Most blind snakes are regarded as egg layers, but this has not been positively determined for this species. Completely harmless and relies on emitting an unpleasant odour from the anal glands to deter predators.

## *Ramphotyphlops ligatus* (Peters, 1879)

**Description** A pale snout, dark greyish-brown to dark purplish-brown dorsally and creamy white to pink below. Midbody scales are in 24 rows. The snout is smoothly rounded when viewed from above and in profile. Nasal cleft visible from above, almost dividing the nasal scale and contacting the labial below. Grows to about 0.5 m. It has a short tail, ending abruptly in a sharp spine. The eye is greatly reduced and is covered by a scale. The mouth is small and covered by a protrusive snout. The body scales and ventrals are equal in size and are highly polished to provide less friction when the snake is moving through the earth.

**Habits and habitat** Fossorial and shelters in ant nests, termite mounds or termitaria, soil beneath leaf litter, rocks and logs. Rarely active on the ground surface, but most commonly encountered after rains during summer nights. Almost exclusively feeds on termites and their eggs, and the pupae and larvae of ants. Most blind snakes are regarded as egg layers, but this has not been positively determined for this species. Completely harmless and relies on emitting an unpleasant odour from the anal glands to deter predators.

# First aid for snakebite

## Avoid the problem of snakebite

Many, indeed the majority of, snakebites occur needlessly. People should never interfere with snakes or try to touch them, and this particularly applies to children. Children should be taught that snakes are dangerous and that they should leave them alone. When bushwalking through long grass, travellers should always wear protective clothing, such as long trousers and boots. A torch should be used when walking about at night time. The areas around houses and playgrounds should be kept clean, rubbish removed and the grass cut.

## First aid for snake bite patient

*'In practical terms, venom movement can be effectively delayed for long periods by the application of a firm "bandage" to the length of the bitten limb, combined with immobilisation by a splint. Pressure alone or immobilisation alone do not delay venom movement.'*

(S. K. Sutherland, 1979)

The optimal first aid management of snake bite is as follows:

1. Apply a firm, broad, pressure bandage to the bitten limb and, if possible, to the length of the limb.
2. The limb should be immobilised by a splint and kept as still as possible.
3. Keep the patient still.

Immobilisation and the use of a pressure bandage reduces the movement of venom from the bite site. This restriction of the venom movement will allow more time to transport the patient to hospital. The victim should keep still and rest. If possible, transport be brought to the patient rather than vice versa.

Incising and excising snakebites is absolutely contra-indicated. There is no evidence that such management is of any use, and it is clearly dangerous, since incising the patient runs the risk of cutting tissues.

Torniquets are not recommended. Indeed, these may be dangerous because they are painful and when they are released there is a surge of blood into the limb, causing reflex hyperaemia (increased blood flow), which results in a rapid central movement of the venom.

The bite site should be left alone and not washed or wiped. At the hospital, the doctor will take a specimen from the bite site, which may help with snake identification. If the snake is dead, it should be taken to hospital to help with identification.

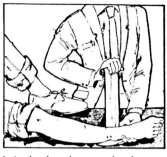

1 Apply a broad pressure bandage over bite site.

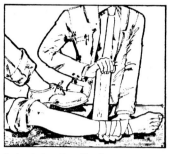

2 Apply bandage firmly.

3 Extend bandage as high as possible.

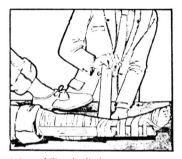

4 Immobilise the limb.

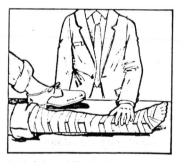

5 Limb immobilised with pressure bandage in place.

# Conservation

Australia is one of the most diverse land masses on earth. It has tropical and temperate forests, woodlands, arid deserts, heathlands, mangroves, shrublands, grasslands and alpine ecosystems. Despite the large land mass and relatively low population density, in 200 years of European settlement, much of the Australian landscape has suffered severe degradation. In that time we have lost as many or more plant species as Europe or North America. In the agricultural regions, over 90 per cent of native vegetation has been lost and erosion and salinity are enormous problems. Our largest river network, the Murray-Darling drainage system, is threatened by salinity.

The specific threats to the ophidian fauna of Australia cannot be divorced from the major environmental problems facing Australia now and in the near future. The World Wildlife Fund has summarised these major threats to the environment. They are:

1  Clearance of native vegetation for agriculture and urban development, particularly in southern Australia.
2  Degradation of native vegetation by pastoralism and altered fire regimes, particularly in northern tropical Australia and the arid zone.
3  Development of tourist facilities and other infrastructure in coastal regions.
4  Over exploitation of natural resources such as timber, soils and fish.
5  The presence of introduced animals and plants in all habitats.
6  The presence of plant diseases, especially the root-rot fungus *Phytophthera* spp.
7  The development of mineral/coal/oil and gas deposits in sensitive sites such as river headwaters and near coral reefs.
8  Lack of expertise and data on fire management in most habitats.
9  Climatic change arising from global air pollution.

In addition to the loss of habitat, the introduction of feral animals in Australia has been devastating to many of our fauna types and has directly led to the extinction of a number of our species. Such animals as foxes, cats, donkeys, horses, rabbits, camels, goats, cane toads, the water buffalo, a number of exotic fish species, cattle, pigs, black rat, brown rat, European mouse, starlings, sparrows, Indian minas, house geckoe and a blind snake species are some of the introduced animals competing and preying on our native species. Most of the animals listed, were deliberately introduced to Australia. In many cases there was a campaign of introduction. However, animals such as the European mouse are established in nearly every country in the world and have been enormously successful travellers and colonisers. In arid regions, feral cats have been particularly detrimental to reptile species, preying on a range of species including venomous snakes. They also prey on mammals and birds, and this may have an indirect effect by decreasing the prey available to native animals, in particular snakes.

In consultation with the state statutory bodies, the Australian National Parks and Wildlife Service has produced a list of endangered species. On this list, Table 1, the summary presents a picture of Australian wildlife as it is today resulting from our short 200 years of settlement.

| Mammals | extinct | 16 |
| | endangered | 26 |
| Birds | extinct | 3 |
| | endangered | 35 |
| Frogs | extinct | nil |
| | endangered | 7 |
| Reptiles | extinct | 1* (not on the ANPWS list. Refers to Adelaide bluetongue *Tiliqua adelaidensis*) |
| | endangered | 9 |
| Fish * | extinct | nil |
| | endangered | 5 |

Table 1   Summary of the status of Australian Wildlife prepared by ANPWS.

This list is arguably conservative. Some state authorities would probably be a little more aggressive in their summary of the species status. Nevertheless, with at least 20 animals extinct and 82 endangered, we have an enormous problem on our hands.

One of the greatest impediments to implementation of conservation measures in Australia, is our system of government. We have no less than 8 state and territory statutory bodies and an additional Federal agency all with various degress of control over different environmental matters. The states and territories are responsible for nearly all matters concerning internal Australian affairs whereas the ANPWS controls all foreign dealings. To implement any conservation strategy with national significance, cooperation between these bodies is imperative. Because of the many bodies involved and the varying pressures on them, total agreement is difficult. There are a number of anti-conservationary directions set by various individual governments. For instance, Queensland has more interest in development than conservation and the competing interests in land use between primary production, tourism development and National Parks are of concern for the conservation lobby. In Western

*Inland taipans*
Oxyuranus
microlepidotus
*mating in captivity at Venom Supplies. Tanunda, South Australia. (photo by P. Mirtschin).*

*Colletts snakes*
Pseudechis colletti,
*exhibiting male combat. This type of behaviour is more likely to be discovered in captivity. (photo by P. Mirtschin).*

Australia, the combination of the forestry and environment departments, two diametrically opposed forces, has produced a service with diluted interests in the environment. Even of more concern is the issue of law enforcement. Some states require import and export documentation when moving wildlife between states and territories, while others have abandoned this control measure. For the system to work effectively, all states must be uniform in their administration of this procedure. To complicate things even further, this practice is in direct conflict with the Australian Constitution which advocates free trade between the states.

189

*The degraded habitat of the Kreffts tiger snake* Notechis ater ater. *(photo. by P. Mirtschin).*

*Habitat of Kreffts tiger snake* Notechis ater ater. *Same creek as above, 200m away and fenced to exclude stock (photo by P. Mirtschin).*

*Gross habitat destruction by clearing Eyre Peninsula South Australia. (photo by P. Mirtschin).*

As a result of the many governments being involved in conservation, the administration of our National Parks and Wildlife service is handicapped. The duplication in each state and territory of various services and the constant need to consult between the states make administration more costly than necessary. Even more serious is the difficulty in getting sensible and meaningful reforms approved in each state.

Australia is a big country, with a relatively small population. Our natural resources will restrict population growth to about 20 to 25 million. Even when this ceiling population is reached, our capacity to raise sufficient revenue from taxation to substantially bolster the coffers of the statutory bodies administering environmental matters, is not going to increase much further than it is today. Our only solution to achieve more from the conservation dollar derived from taxation is to become more efficient. It has been said that 'Australians would be better served by the elimination of the second tier of government — that is the states — which no longer serve their original purpose and act as a positive impediment to achieving good government'. (Bob Hawke 1979 — Resolution of Conflict). This very simple idea, with enormous implications, would not only make our country far more efficient in many areas, it would also improve the chances of survival of the many threatened species in Australia today by making more funds available to spend on threatened species rather than in unproductive bureaucracies. If environmental matters were all handed back to the Commonwealth, there would only be one government body to finance instead of 9. It should be viewed as reverting to the most natural way of handling conservation, for many of our fauna occur in a number of states and territories, none of the pest species threatening wildlife observe state boundaries and implementation of uniform reforms would be easier.

The many macroscopic environmental pressures have been outlined above. Estimates for the annual mortality rates for reptiles through most causes are as follows in Table 2:

| **Natural** | | |
|---|---|---|
| Predation, disasters, disease, and ageing | 3,840,000,000 | 99.77% |
| **Human induced** | | |
| Road kills | 5,480,000 | 0.14% |
| Mallee clearing | 3,260,000 | 0.08% |
| Collecting for research | 20,000 | 0.0005% |
| Seasnake skins, amateur collecting | <20,000 | 0.0005% |

Table 2   Estimates of annual mortality taken from Ehmann and Cogger 1985.

191

*Winter and summer colour variations of the inland taipan* Oxyuranus microlepidotus. *Seasonal colour variations are more easily discovered in captivity. (photo by P. Mirtschin).*

*Clutch of death adders* Acanthophis antarcticus. *Captive bred clutches can contribute to understanding of the species. (photo by P. Mirtschin).*

From the figures in Table 2, it can be seen that collecting for research or amateur collecting accounts for only a small percentage of the total annual cull of all reptiles. Amazingly and disproportionately, much of the effort of the statutory bodies surrounds controlling the taking and keeping of reptiles. There have been numerous court appearances around the country which 'have almost entirely involved what are best described as petty infringements' (Ehmann and Cogger 1985). Given the macroscopic problems facing conservation today, it seems quite out of order to be pursuing policies to control activities that have little bearing on the total cull of most reptile populations. On the other hand, the efforts in this direction could arguably be said to have a negative effect on reptile conservation. Given the environmental problems facing wild animals, it is inevitable that many species will face extinction unless captive breeding is mounted. The protection agencies are unlikely to have the funds for initiating any significant projects in this direction since they are already short of funding to manage their everyday activities. The only real hope is for those private individuals with the skills to be encouraged to participate in such programs with identified species. Such a program immediately faces difficulties when sustainability is considered. The very laws discourage newcomers to learn the skills, especially when they are continually harassed by the law enforcement agencies over trivial matters like book keeping, preserved animals off permit and collecting technicalities. One estimate considers that in 1973 there were 4,000 reptile keepers in Australia, and by 1980 there were less than 1000 (Hoser 1989). The 1973 figure is probably overestimated and 2000 is probably nearer an accurate figure. The 1980 figure of less than 1000 is consistent with Ehman and Cogger 1985 (982 keeping amateurs). No matter which figures are accepted, there has been a dramatic reduction in amateur herpetologists in line with the more stringent regulations. Since most of us learn more effectively by doing rather than hearing or seeing things, hands on involvement with reptiles is the only way proper education will continue. The incentive to learn has been removed. Hands on keeping is also the only practical way to learn about behavioural aspects of certain species.

Some protection agencies openly admit that they want to reduce the number of reptile keepers. Western Australia has suppressed herpetology ever since they passed laws protecting reptiles. It is little wonder that behavioural studies, captive breeding, and research conducted by interested naturalists is at a minimum from that state. In fact, the lack of disseminated information emanating from Western Australia, is a situation best described as a literary desert. The other states are not much better, for they all indulge in the most trivial of exercises, at great cost to the taxpayer, and achieving nothing but the alienation of many who would be, in most cases, champions of the cause for reptile conservation. When asked to justify their stand, their arguments will never be on a scientific basis, and inevitably 'Yes Minister' type reasons will be given. For those cavaliers who decide they are going to change the system, they will be led on a merry course from this politician to that public servant, maybe to a government committee and even from one government to another and in the final analysis will have achieved nothing. The inevitable result will be that the timely advice, query or complaint from the enthusiastic herpetologist will be forgotten, the system grinds on and the herpetologist gives up and takes up golf.

From a scientific viewpoint, we need those interested herpetologists. Conservation of reptiles, more than ever, depends on a dramatic rethinking of the protection laws by the statutory bodies. If there aren't any changes, one could quite legitimately accuse the statutory bodies of actively allowing some species to become threatened and extinct. This will happen because we haven't got sufficient people out monitoring their status, and the law enforcement agencies are mismanaging their time by pursuing matters relating to the wider wildlife spectrum instead of targeting sustainable preservation of important endangered species. An example in South Australia serves to illustrate the point. Most of the law enforcement resources are put into catching offenders for having slight discrepancies with reptiles kept under permit, for importing or exporting between the states without permits or taking some common reptiles without a permit. These offences are reported to the public as major crimes, but put in perspective, they have no bearing on the sustainable conservation future of the species concerned. Recently, a large number of Kreffts tiger snakes *Notechis ater ater* from the southern Flinders Ranges, which are a threatened species, were taken illegally for taxodermy. If more attention were given to endangered species, such as regular patrols of their habitats, events such as this would be minimised. Even more alarming is the fact that officers engaged in caring for endangered species in South Australia at the captive breeding site at Monarto, have to spend most of their time caring for confiscated common reptiles instead of devoting all that time to caring for endangered species. Blanket protection has many hidden disadvantages.

There is a mounting feeling both among professional zoologists and amateur herpetologists that the wildlife laws, as they pertain to reptiles, require a major overhaul. So much so that the First World Congress of Herpetology held in Canterbury in Kent U.K. September 1989 established a committee to promote the recognition of amateur herpetology worldwide. The committee has been named the Committee on the preservation of Reptiles and Amphibians through captive Husbandry and Propagation (CPRACHP). The draft objectives drawn up were as follows:

1. To establish that the live animal trade, as used by pet keepers, zoo keepers, and professional herpetologists, is a negligible factor in population depletion;

2. To admit that most herpetologists gained their interest and much of their training with 'pet' reptiles and amphibians.
3. Acknowledge that the subsequent contributions to herpetology by amateurs has been considerable. It includes entry into the profession by some, serious husbandry and breeding efforts by others, and serious conservation efforts by others.
4. To demonstrate that the current conservation laws are contrary to the needs of species preservation, in that they
   a. are written for ease of enforcement, which severely limits access to specimens by qualified students;
   b. provide protection to animals but NOT their habitats, which effectively renders them useless;
   c. make serious amateurs and many professionals criminals in the eyes of the law, or make them near enough to criminal to elicit ill-will from many colleagues.
5. To move herpetological societies with large or predominantly professional memberships to acknowledge as a matter of policy that
   a. contributions made by amateurs are of importance;
   b. such serious amateur efforts should be encouraged by co-operation from professionals, and;
   c. laws restricting access to animals need serious revision that will allow greater access to live specimens by all interested persons.

The fallacy that blanket protection laws are effective in maintaining viable populations of reptile species is demonstrated when one examines the inconsistency of species protection around Australia. Amphibians are not protected in a number of states; yet there is no evidence to suggest they are worse off in those states as a result or, conversely, better off in the states where protection is afforded. A case could be made, however, to suggest that those states that do not protect frogs are enlightened and more is being achieved in those states for amphibian conservation. Even more contradictory is the fact that Tasmania does not protect snakes and again there is no evidence suggesting that snakes are declining in that state as a result.

To gauge the feeling across Australia, it is useful to summarise the opinions of a number of reptile experts around Australia. Those opinions have been divided into a number of specific areas:

# *Self-education, and the need for education*

'Many an ardent conservationist has started off keeping animals as pets and progressed from watching them in captivity to making field observations. The salient point is that having first come to know them from observations in captivity, his interest was sufficiently aroused to make the much harder step of watching them at liberty in nature'. (Bustard 1970)

'The knowledge of wildlife officers in most states is deficient in reptiles'. (Mirtschin and Davis 1982)

# *Disadvantages of blanket legislation and arguments for its removal*

'I know of absolutely no grounds to justify or necessitate legislation to control the taking of frogs anywhere in Australia except for the handful of endangered species'. (M. Tyler 1979).

'. . . Australian wildlife legislation: in most cases it is primarily aimed at controlling the collecting of specimens, and Tyler's comment' (1979 — above) 'is equally valid when applied to reptiles'. (Rawlinson 1980).

'Sometimes it is argued that those involved in law enforcement might be unable to distinguish between the species genuinely in need of protection and those not requiring it. This may be the case, but it conjures the analogy of putting every single Australian person in prison, on the quite legitimate argument that police are unable to distinguish innocent citizens from the few criminals in their midst'. (M. Tyler 1979).

'With the possible exception of crocodiles and some marine turtles in northern Australia, and perhaps some local populations of lizards, frogs and tortoises in the south, reptile and frog populations in Australia are rarely endangered by direct human exploitation of the animals as a resource, ie. by hunting for skins or for pleasure, or being captured alive for the local pet or education market'. (Cogger 1975).

'The new N.P.W.S. policy on permits is more for the sake of administration than reptile welfare'. (T. Houston 1976 Prev. S.A. Museum curator of reptiles).

'There is widespread concern about the effects that protective legislation (and, more so, the way it is being implemented) is having on herpetology in Australia'. (Herpetofauna 1976).

The existing wildlife regulations 'tend to discourage amateur herpetologists and students, and alienates many of those who would otherwise be strong supporters of long-term conservation measures'. (Ehmann and Cogger 1985).

'There is no biological evidence to indicate that normal herpetological activities have led to declines in reptile populations'. (Herpetofauna 1976).

'Alteration of habitat was recognised as the major cause of decline in wildlife populations; of far greater significance than direct exploitation of wildlife by man'. [Fox Report — Parks & Wildlife Dec 1974 1 (4).]

Export of Australian fauna is prohibited. 'Their consequent rarity overseas has led to smuggled specimens commanding artificially high prices on international markets'. (Ehmann and Cogger 1985).

## *Indicators of what should be done*

A statement regarding the numbers of animals permanently removed from the wild 'has no significance unless it is related to the total number of animals in the population and their rate of replacement'. (Frith 1973).

'One objective of conservation is to maintain maximum genetic diversity'. (Jenkins 1985).

## *Support for specific legislation and a unified approach with less government.*

'. . . reptile and frog species/populations at risk can be realistically identified'. (Ehmann and Cogger 1985).

'Each state and Territory has enacted wildlife conservation legislation and has a body to administer it . . . These vary widely in stature' . . . 'There are many conservation problems that affect more than one state and joint solutions have to be sought . . . There is a strong case for an increased Commonwealth participation in projects that are national in scope'. (Costin and Frith 1971).

'. . . the present bias towards protection of individual specimens obscures the needs of species-as-populations'. (Ehmann and Cogger 1985).

'The most important thing in wildlife conservation is to recognize the targets and the animals most needful of action'. (Costin and Frith 1971).

'There are deficiencies in resource conservation in Australia'. (Jenkins 1985).

Duplication of authority has resulted in parochial attitudes towards wildlife conservation'. (Jenkins 1985).

'The economics of conservation are such that governments must necessarily allocate priorities'. (Jenkins 1965).

Legal protection of frogs and reptiles 'has been sequential and without uniformity. However protection of land as national parks or equivalent reserves automatically protects the herpetofauna occurring on that land'. (Jenkins 1985).

# *Support for more research*

'A great deal of research remains to be done, since the full habits of virtually all species are to date unknown'. (Gow 1976).

The species we consider threatened are the Adelaide copperhead *Austrelaps superbus* (pygmy group), the Kreffts tiger snake *Notechis ater ater* and the Broad headed snake *Hoplocephalus bungaroides*. Other species that should be included as threatened for the time being, based on lack of knowledge are: the rough scaled python *Morelia carinata* and two small elapids *Echiopsis atriceps*, *simoselaps mimimus* and one colubrid *Stegnotus parvis*.

Five sea snakes should also be included purely because we know very little about their population status in Australian waters. They are: *Aipysurus tenuis*, *Hydrophis atriceps*, *Hydrophis belcheri*, *Hydrophis inornatus* and *Hydrophis vorisi*.

As can be seen from the arguments above and the widespread opinion, there is a well supported plea for major revision in the way Australia handles its wildlife conservation. It is our belief that the best way to manage wildlife conservation in Australia would be to hand all these responsibilities to the Commonwealth and for the Commonwealth to completely redraft all legislation so that the effort is concentrated where it is needed most. In the past the recommendations of reptile experts have been largely ignored or at least the various recommending organisations have been forced to accept as inevitable, the laws wanted by the authorities. Australia can't afford to continue managing its wildlife under the hopelessly inefficient system it has now. Not only is it costing the taxpayer more, but because of the waste in duplication of responsibilities, the cost of managing blanket legislation with little benefit to the wildlife, less can be spent on endangered species. This leads us to the conclusion that this system is the direct cause of greater rates of extinctions and species driven to threatened status than would be experienced under a more efficient system of government with specific legislation.

# Further reading

ARCHER, M. (1972). 'Domestic cats and dogs — a danger to the Australian fauna', *W.A. Nat.*, 85–87.

BARNETT, B. & SCHWANER, T. (1985). 'Growth in captive born tiger snakes (*Notechis ater serventyi*) from Chappell Island: Implications for field and laboratory studies', Trans. Roy. Soc. S. Aust., 109: 7.

BAYLY, C.P. (1976). 'Observations on the food of the feral cat (*felis catus*) in an arid environment', *Sth. Aust. Nat.*, 51 (2): 22–24.

BROAD, A. J., SUTHERLAND, S. K. & COULTER, A. R. (1979). 'The lethality in mice of dangerous Australian and other snake venoms', *Toxicon*, 17: 664–67.

BUSTARD, R. H. (1970). *Australian Lizards*, Collins, Brisbane.

CANN, J. (1986). *Snakes Alive*, Kangaroo Press, Sydney.

COGGER, H. G. (1983). *Reptiles and Amphibians of Australia*, Reed, Sydney.

COSTIN, A. B. & FRITH, H. J. (1971). *Conservation*, Penguin Books, Melbourne.

COVACEVICH, J. & ARCHER, M. (1975). 'The distribution of the cane toad, *Bufo marinus*, in Australia and its effects on indigenous vertebrates', Mem. Qld. Mus., 17 (2): 305–10.

COVACEVICH, J. & WOMBEY, J. (1976). 'Recognition of *Parademansia microlepidotus* (McCoy) (Elapidae) a dangerous Australian snake', Proc. Roy. Soc. Qld., 87: 29–32.

DAVIS, R. & MIRTSCHIN, P. J. (1985). 'Snake Bites: A Guide to Treatment', *Patient Management*, 91–99, September.

Draft World Wildlife Fund Australia Conservation Programme 1990–1992, April 1990.

EHMANN, H. & COGGER, H. G. (1985). 'Australia's endangered Herpetofauna: A review of criteria and policies', in *Biology of Australasian Frogs and Reptiles*, eds G. Grigg, R. Shine & H. Ehmann, Roy. Zoo. Soc. of NSW, pp 435–47.

FRITH, H. J. (1973). *Wildlife Conservation*, Angus & Robertson, Sydney.

GOPALAKRISHNAKONE, P. (1985). 'Structure of the spinous scales of *Lapemis hardwickii* (1) Light, transmission electron and scanning electron microscope study', *The Snake*, vol. 17, 148–55.

GOW, G. F. (1976). *Snakes of Australia*, Angus & Robertson, Sydney.

GOW, G. F. (1989). *Graeme Gow's complete guide to Australian snakes*, Angus & Robertson, Sydney.

HEATWOLE, H. (1987). *Sea Snakes*, New South Wales University Press, Sydney.

HOSER, R. (1989). *Australian Reptiles and Frogs*, Pierson & Co., Sydney.

JENKINS, R. W. G. (1985). 'Government legislation and conservation of endangered reptiles and amphibians', in *Biology of Australasian Frogs and Reptiles*, eds. G. Grigg, R. Shine & H. Ehmann, 431–33.

KAY, P., GARLEPP, M. & DAWKINS, R. L. (1979). 'Evidence for human anti-venom antibodies and detection of anti-acetylcholine receptor antibodies using dugite (*Pseudonaja affinis*) venom', *Neurotoxins & Clinical Advances*, eds I. W. Chubb & L. B. Geffen, Centre for Neuroscience, Flinders Medical Centre.

LONGMORE, R. (1986). *Atlas of elapid snakes of Australia*, Aust. Govt. Publishing Service, Canberra.

MENGDEN, G. A., SHINE, R. & MORITZ, C. (1986). 'Phylogenetic relationships within the Australian venomous snakes of the genus *Pseudechis*', *Herpetologica*, 42 (2): 215–29.

MIRTSCHIN, P. J. & DAVIS, R. (1982). *Dangerous Snakes of Australia*, Rigby, Adelaide.

MIRTSCHIN, P. J. (1985). 'An overview of captive breeding of common death adders, *Acanthophis antarcticus*, (Shaw) and its role in conservation', in *Biology of Australasian Frogs and Reptiles*, eds. G. Grigg, R. Shine & H. Ehmann', Roy. Zoo. Soc. of NSW.

MIRTSCHIN, P. J., CROWE, G. R. & THOMAS, M. W. (1984). 'Envenomation by the inland taipan *Oxyuranus microlepidotus*', *Med. J. Aust.*, 141: 850–51.

MIRTSCHIN, P. J. (1982). 'Seasonal Colour Changes in the Inland Taipan *Oxyuranus microlepidotus* (McCoy)', *Herpetofauna*, 14 (2): 97–99.

MIRTSCHIN, P. J. & REID, R. B. (1982). 'Occurrence and Distribution of the Inland Taipan *Oxyuranus microlepidotus* (Reptilia: Elapidae) in South Australia', (brief communication) Trans. Roy. Soc. of S.A., 106: (4): 213–14.

MIRTSCHIN, P. J. (1983). 'The Common Death Adder *Acanthophis antarcticus* in South Australia: Its Status and Conservation', *Sth. Aust. Nat.*, 58 (2): 24–28.

MIRTSCHIN, P. J. (1982). 'Further Notes on Breeding Death Adders (*Acanthophis antarcticus*) in Captivity', *Herpetofauna*, 13 (2): 14–17.

MIRTSCHIN, P. J. (1981). 'South Australian Records of the Inland Taipan (*Oxyuranus microlepidotus*) (McCoy, 1879)', *Herpetofauna*, 13 (1): 20–23.

MIRTSCHIN, P. J. (1981). 'Black Tiger Snakes from the Sir Joseph Banks Islands', *Sth. Aust. Nat.*, 55 (4): 58–61.

MIRTSCHIN, P. J. (1988). 'Double Egg Laying in Taipans *Oxyuranus scutellatus*', 10th International Conference on Captive Propagation and Husbandry, San Antonio, Texas, 149–57.

MIRTSCHIN, P. J. (1988). 'Captive Breeding in the king brown snake *Pseudechis australis* from Eyre Peninsula', 10th International Conference on Captive Propagation and Husbandry, San Antonio, Texas, 141–48.

MIRTSCHIN, P. J. & BAILEY, N. (1990). 'Study of the Kreffts Black Tiger Snake *Notechis ater ater* (Reptilia: Elapidae)', *S.A. Nat.* 64 (3/4): 52–61.

MIRTSCHIN, P. J., CROWE, G. R. & DAVIS, R. (1990). Dangerous Snakes of Australia. In eds P. Gopalakrishnakone and L. M. Chou. Snakes of Medical Importance (Asia–Pacific Region). Venom and Toxin Research Group. National University Singapore and International Society of Toxinology (Asia-Pacific Region).

MIRTSCHIN, P. J., HUDSON, B. & CURRIE, B. A monograph on *Acanthophis* (Death Adders) in Australia and New Guinea for World Heath Organisation, IST, (in press).

OVINGTON, D. (1978). *Australian Endangered Species*, Cassell Australia, Sydney.

Report on First World Congress of Herpetology. *Herp. News*. Australian Herpetological Society and Reptile Keepers Association, Issue 9 Spring 89/Summer 90, 11–13.

SCHWANER, T. (1985). 'Population structure of black tiger snakes, *Notechis ater niger*, on offshore islands of South Australia', in *Biology of Australasian Frogs and Reptiles*, eds. G. Grigg, R. Shine & H. Ehmann, Surrey: Beatty & Sons Pty Ltd., Roy. Zoo. Soc. of NSW.

SHINE, R. & COVACEVICH, J. (1983). 'Ecology of Highly Venomous Snakes: The Australian Genus *Oxyuranus* (Elapidae)', *Journal of Herpetology*, 17 (1): 60–69.

SHINE, R. (1987). 'Ecological ramifications of prey size: Food habits and reproductive biology of Australian copperhead snakes (*Austrelaps*, Elapidae)', *Journal of Herpetology*, 21 (1): 21–28.

SHINE, R. (1980). 'Reproduction, feeding and growth in the Australian burrowing snake *Vermicella annulata*', *Journal of Herpetology*, 14 (1): 71–77.

SHINE, R. (1980). 'Comparative ecology of three Australian snake species of the genus *Cacophis* (Serpentes: Elapidae)', *Copeia*, (4): 831–38.

SHINE, R. (1980). 'Ecology of eastern Australian whipsnakes of the genus *Demansia*', *Journal of Herpetology*, 14 (4): 381–89.

SHINE, R. (1981). 'Venomous snakes in cold climates: Ecology of the Australian genus *Drysdalia* (Serpentes: Elapidae)', *Copeia*, (1): 14–25.

SHINE, R. (1986). 'Natural history of two monotypic snake genera of southwestern Australia, *Elapognathus* and *Rhinoplocephalus* (Elapidae)', *Journal of Herpetology*, 20 (3): 436–39.

SHINE, R. (1989). 'Constraints, allometry and adaptation: Food habitats and reproductive biology of Australian brown snakes (*Pseudonaja*: Elapidae)', *Herpetologica*, 45 (2): 195–207.

SHINE, R. & FITZGERALD, M. (1989). 'Conservation and reproduction of an endangered species — the broad-headed snake, *Hoplocephalus bungaroides* (Elapidae)', *Aust. Zoologist*, vol. 25 (3): 65.

SHINE, R. (1984). 'Ecology of small fossorial Australian snakes of the genera *Neelaps* and *Simoselaps* (Serpentes, Elapidae)', in *Vertebrates ecology and systematics — A tribute to Henry S. Fitch*, edited by R. A. Seigel, L. E. Hunt, J. L. Knight, L. Maluret & N. L. Zuschlag, Mus. Nat. Hist., University of Kansas, Lawrence, 173–183.

SHINE, R. (1980). 'Ecology of the Australian death adder *Acanthophis antarcticus* (Elapidae): Evidence for convergence with the Viperidae', *Herpetologica*, 36 (4): 281–89.

SHINE, R. (1983). 'Food habits and reproductive biology of Australian elapid snakes of the genus *Denisonia*', *Journal of Herpetology*, 17 (2): 171–75.

SHINE, R. (1982). 'Ecology of the Australian elapid snake *Echiopsis curta*', *Journal of Herpetology*, 16 (4): 388–93.

SHINE, R. (1981).'Ecology of Australian elapid snakes of the genera *Furina* and *Glyphodon*', *Journal of Herpetology*, 15 (2): 219–24.

SHINE, R. (1987). 'Food, habits and reproductive biology of Australian snakes of the genus *Hemiaspis* (Elapidae)', *Journal of Herpetology*, 21 (1): 71–74.

SHINE, R. & SCHWANER, T. (1984). 'Prey constriction by venomous snakes: A review, and new data on Australian species', *Copeia*, (4): 1067–71.

SHINE, R. (1987). 'Ecological comparisons of island and mainland populations of Australian tiger snakes (*Notechis*, Elapidae)', *Herpetologica*, 43 (2): 233–40.

SHINE, R. (1983). 'Arboreality in snakes: Ecology of the Australian elapid genus *Hoplocephalus*', *Copeia*, (1): 198–205.

SHINE, R. & CHARLES, N. (1982). 'Ecology of the Australian elapid snake *Tropidechis carinatus*', *Journal of Herpetology*, 16 (4): 383–87.

SHINE, R. (1984). 'Reproductive biology and food habits of the Australian elapid snakes of the genus *Cryptophis*', *Journal of Herpetology*, 18 (1): 33–39.

STORR, G. M. (1978). 'Whip snakes (*Demansia*, Elapidae) of Western Australia', Rec. West. Aust. Mus., 6 (3): 287–301.

STORR, G. M. (1981). 'The genus *Ramphotyphlops* (Serpentes: Typhlopidae) in Western Australia', Rec. West. Aust. Mus., 9 (3): 235–71.

SUTHERLAND, S. K. (1983). *Australian Animal Toxins*, Oxford University Press, Melbourne.

*Toxic Plants and Animals*, (1987), eds Jeanette Covacevich, Peter Davie & John Pearn, Qld. Museum.

TYLER, M. J. (1979). 'The impact of European man upon Australasian amphibians', in *The status of endangered Australasian wildlife*, ed. Michael J. Tyler, proceedings of the Centenary Symposium of the Royal Zoological Society of South Australia, Adelaide, 21–23 Sept. 1978.

WILSON, S. K. & KNOWLES, D. G. (1988). *Australia's Reptiles: A photographic reference to the terrestrial reptiles of Australia*, Collins, Sydney.

WORRELL, E. (1963). *Reptiles of Australia*, Angus & Robertson, Sydney.

# Glossary

| | |
|---|---|
| aboreal | living in trees |
| anal scale | ventral scale just anterior of anus |
| anterior | head end of snake |
| anticoagulant | inhibits or prevents blood clotting |
| asphyxiation | death caused by suffocation |
| bifid | divided by deep cleft into 2 parts |
| buccal cavity | the mouth cavity |
| caudal | relating to the tail |
| chromosome | a nuclear body of DNA and protein, comprised of a linear sequence of genes |
| cloacal spur | vestigial spurs on either side of the vent |
| coagulant | causes blood to clot |
| crepuscular | active at the transition between day and night |
| cytogenetic | chromosomal mechanisms, their behaviour and effects on inheritance and evolution |
| diurnal | active during the day |
| dorsal | upper surface of snake |
| elapid | fixed front fanged snake |
| electrophoretic patterns | separation of toxins in a neutral viscous medium by the influence of an electric field |
| foliform | leaf shaped with extensive free edge |
| fossorial | adapted for digging or burrowing into ground |
| frontal scale | a large median scale between the eyes |
| haemolytic | toxin causing damage to blood cells |
| haemorragic | toxin causing damage to blood vessels |
| hyperaemia | increased blood flow |
| infralabial scales | scales on bottom lips |
| imbricate | overlapping |
| labial scales | scales bordering lips |
| lanceolate | shaped like the head of a spear |
| lingual fossa | a notch on the bottom of the rostral scale to allow the protrusion of the tongue |
| littoral | shoreline |
| morphology | form and structure of snakes |
| myasthenia gravis | an inherited muscular debility |
| myotoxin | muscle destroying toxin |
| nape | back of neck |
| neonate | recently hatched or born specimen |
| neurotoxin | toxin affecting the nervous system |
| nocturnal | active at night |
| nuchal | pertaining to the nape |
| ophidian | Snakes |

| | |
|---|---|
| oviparous | egg laying |
| oviposition | when the snake deposits eggs or egg sacs |
| ovoviviparus | eggs that hatch during or soon after oviposition |
| parasitisation | infested with parasites |
| posterior | tail end of snake |
| postocular scales | scales behind the eyes |
| postsynaptic | the receiving surface in the neuromuscular junction |
| prehensile | capable of clasping |
| preocular scale | scales along front margin of eye |
| presynaptic | acting just prior to the nerve ending at the neuromuscular junction. The transmission surface |
| procoagulant | a coagulant, toxin causing coagulation |
| reticulum | a network |
| rostral scale | scale(s) on the nose tip |
| rugose | a wrinkled or uneven surface |
| sclerophyl | hard stiff leaf (as in eucalypts) |
| subcaudal | under surface of tail |
| sublingual gland | gland under tongue for salt excretion |
| supra (prefix) | upper |
| supralabials | scales on upper lips |
| supraocular scales | scales above the eyes |
| synapse | the nerve ending at the neuromuscular junction |
| temperate | moderate, lacking extremes |
| termitaria | termite mounds, complexes or nests |
| terrestrial | occurs on or is active on the ground |
| thrombocyto-penia | damage to blood platelets |
| tubercle | small rounded protuberance on the skin |
| valvular nostrils | nostrils with valves to prevent inflow of water |
| viper | snake possessing rotating fangs |
| venom | poison |
| ventral | under surface of snake |
| vertebral | along the spine |
| vestigial | remnant |
| viviparous | live bearing |

# Index of Common Names

# Index of Scientific Names

# List of Photographs

## Typhlopidae: Blind Snakes

*Ramphotyphlops affinis*          S. Wilson (head shot), *181*
*Ramphotyphlops australis*       S. Wilson, *182*
*Ramphotyphlops grypus*         S. Wilson, *183*
*Ramphotyphlops ligatus*        P. Mirtschin, *184*

## Boidae: Pythons

*Aspidites melanocephalus*       P. Mirtschin, *16*
*Aspidites ramsayi*               P. Mirtschin, *17*
*Chondropython viridus*         P. Mirtschin, *19*
*Liasis albertisii*               P. Mirtschin, *20*
*Liasis childreni*                P. Mirtschin, *21*
*Liasis fuscus*                   P. Mirtschin, *22*
*Liasis olivaceus*                P. Mirtschin, *23*
*Liasis perthensis*              S. Wilson, *24*
*Morelia amethistina*          P. Mirtschin, *25*
*Morelia bredli*                  G. Harold, *26*
*Morelia oenpelliensis*        G. Harold, *28*
*Morelia spilota spilotes*       P. Mirtschin, *29*
*Morelia spilota variegata*     P. Mirtschin, *30*
*Morelia spilota imbricus*      G. Harold, *30*

## Acrochordidae: File Snakes

*Arochordus arafurae*          C. Banks, *178*
*Arochordus granulatus*        P. Horner, *179*

## Colubridae: Fangless or Rear Fanged Snakes

*Boiga irregularis*              P. Mirtschin, *166*
*Cerberus rynchops*           S. Wilson, *167*
*Dendrelaphis calligaster*      J. Weigel, *168*
*Dendrelaphis punctulatus*    P. Mirtschin, *169*
*Enhydris polylepis*           P. Horner, *170*
*Fordonia leucobalia*          P. Horner, *172*
*Myron richardsonii*          P. Horner, *173*
*Stegnotus cucullatus*         P. Mirtschin, *174*
*Styporrhynchus (Amphiesma) mairii*    S. Wilson, *176*

## Elapidae: Front Fanged Snakes

| | |
|---|---|
| *Acanthophis antarcticus* | P. Mirtschin, *32* |
| *Acanthophis (sp)* | P. Mirtschin, *33* |
| *Acanthophis (sp)* | D. Robinson, *34* |
| *Acanthophis praelongus* | P. Mirtschin, *35* |
| *Acanthophis pyrrhus* | P. Mirtschin, *36* |
| *Austrelaps superbus* | P. Mirtschin, *37* |
| *A. superbus Pygmy* | P. Mirtschin, *38* |
| *Cacophis harriettae* | S. Wilson, *39* |
| *Cacophis krefftii* | S. Wilson, *40* |
| *Cacophis squamulosus* | S. Wilson, *41* |
| *Crytophis nigrescens* | P. Mirtschin, *42* |
| *Cryptophis pallidiceps* | P. Horner, *43* |
| *Demansia atra* | P. Mirtschin, *44* |
| *Demansia olivacea olivacea* | J. Weigel, *45* |
| *Demansia olivacea calodera* | J. Weigel, *46* |
| *Demansia olivacea rufescens* | G. Harold, *46* |
| *Demansia papuensis papuensis* | J. Weigel, *47* |
| *melaena* | L. Naylor, *47* |
| *Demansia psammophis psammophis* | P. Mirtschin, *49* |
| *cupreiceps* | P. Mirtschin, *50* |
| *reticulata* | H. Cogger, *50* |
| *Demansia torquata* | S. Wilson, *52* |
| *Denisonia devisii* | G. Harold, *53* |
| *Denisonia fasciata* | S. Wilson, *54* |
| *Denisonia maculata* | H. Cogger, *55* |
| *Denisonia (Rhinoplocephalus) punctata* | G. Harold, *56* |
| *Drysdalia coronata* | P. Mirtschin, *57* |
| *Drysdalia coronoides* | S. Wilson, *58* |
| *Drysdalia masteri* | P. Mirtschin, *59* |
| *Drysdalia rhodogaster* | S. Wilson, *60* |
| *Echiopsis atriceps* | S. Wilson, *61* |
| *Echiopsis curta* | P. Mirtschin, *62* |
| *Elapognathus minor* | S. Wilson, *63* |
| *Furina diadema* | P. Mirtschin, *64* |
| *Furina ornata* | G. Harold, *65* |
| *Glyphodon barnadi* | S. Wilson, *66* |
| *Glyphodon dunmalli* | S. Wilson, *67* |
| *Glyphodon tristis* | S. Wilson, *68* |
| *Hemiaspis damelii* | P. Mirtschin, *69* |
| *Hemiaspis signata* | S. Wilson, *70* |
| *Hoplocephalus bungaroides* | P. Mirtschin, *72* |
| *Hoplocephalus bitorquatus* | P. Mirtschin, *71* |
| *Hoplocephalus stephensi* | P. Mirtschin, *74* |

| | |
|---|---|
| *Unechis gouldii* | G. Harold, *125* |
| *Unechis (Rhinoplocephalus) monachus* | G. Harold, *126* |
| *Unechis nigrostriatus* | L. Naylor, *127* |
| *Vermicella annulata annulata* | P. Mirtschin, *128* |
| *Vermicella multifasciata* | P. Horner, *130* |

## Hydrophiidae and Laticaudidae Sea Snakes

| | |
|---|---|
| *Acalyptophis peronii* | H. Cogger, *132* |
| *Aipysurus apraefrontalis* | H. Cogger, *133* |
| *Aipysurus duboisii* | H. Cogger, *134* |
| *Aipysurus eydouxii* | H. Cogger, *135* |
| *Aipysurus foliosquamama* | H. Cogger, *136* |
| *Aipysurus fuscus* | H. Cogger, *137* |
| *Aipysurus laevis* | H. Cogger, *138* |
| *Astrotia stokesii* | P. Horner, *140* |
| *Disteira kingii* | H. Cogger, *141* |
| *Emydocephalus annulatus* | H. Cogger, *143* |
| *Enhydrina schistosa* | H. Cogger, *144* |
| *Hydrelaps darwiniensis* | P. Horner, *146* |
| *Hydrophis coggeri* (juvenile) | H. Cogger, *150* |
| *Hydrophis elegans* | G. Harold, *152* |
| *Hydrophis ornatus* | H. Cogger, *157* |
| *Lapemis hardwickii* | L. Naylor, *160* |
| *Parahydrophis mertoni* | H. Cogger, *161* |
| *Palamis platurus* | H. Cogger, *162* |
| *Laticauda colubrina* | P. Mirtschin, *163* |
| *Laticauda laticauda* | C. Banks, *164* |